North American Jewish Studies

Series Editor

IRA ROBINSON (Concordia University)

ACADEMIC STUDIES PRESS

Anti-*Shechita* Prosecutions in the Anglo-American World, 1855–1913

"A major attack on Jewish freedoms…"[1]

DAVID
FRASER

Boston
2018

1 David Rome, *The Jewish Biography of Henri Bourassa, Part 1* (Montreal: National Archives of the Canadian Jewish Congress, 1988), 94.

Library of Congress Cataloging-in-Publication Data

Names: Fraser, David, 1953-, author.

Title: Anti-shechita prosecutions in the Anglo-American world, 1855-1913: a major attack on Jewish freedoms / David Fraser.

Description: Brighton, MA : Academic Studies Press, [2018] | Series: North American Jewish studies

Identifiers: LCCN 2018006489 (print) | LCCN 2018006683 (ebook) | ISBN 9781618117434 (e-book) | ISBN 9781618117427 (hardback)

Subjects: LCSH: Shehitah—Case studies. | Livestock—Stunning—Case studies. | Animal rights movement—Case studies. | Meat industry and trade—Law and legislation—Case studies. | Religious discrimination—Law and legislation—Case studies. | BISAC: NATURE / Animal Rights. | RELIGION / Judaism / Rituals & Practice. | COOKING / Regional & Ethnic / Jewish & Kosher.

Classification: LCC BM720.S62 (ebook) | LCC BM720.S62 F73 2018 (print) | DDC 296.7/3—dc23

LC record available at https://lccn.loc.gov/2018006489

ISBN (hardback) 978-1-61811-742-7
ISBN (electronic) 978-1-61811-743-4

Book design by Kryon Publishing Services (P) Ltd.
www.kryonpublishing.com

Cover design by Ivan Grave

Published by Academic Studies Press in 2018
28 Montfern Avenue
Brighton, MA 02135, USA
press@academicstudiespress.com
www.academicstudiespress.com

Table of Contents

Acknowledgments

Thanks to the many archivists and librarians who made this book possible. Staff at the Boston Public Library, the British Library, the Halifax Public Library, the London Metropolitan Archives, the Liverpool Record Office, the Manchester Central Library Archives Service, the New York Public Library, the Nova Scotia Archives, and the Wirral Archives Service assisted. Colin Spanjar granted access to the files of the Board of Deputies of British Jews, and Charles Tucker, Record Keeper of the London Beth Din, allowed consultation of the files of the Office the Chief Rabbi. Judi Garner at the American Jewish Historical Society (AJHS) in Boston granted access to the Friedman Family papers. The AJHS in New York provided important information. Rachel Reddick of the Stephen H. Hart Library and Research Center, History Colorado, made available historic Colorado newspaper coverage. Archivist Heather Perez at the Atlantic City Free Public Library assisted in finding local newspaper files, as did staff at the Attleboro and Providence Public Libraries.

Robert Thornton, Senior Archivist, Adelaide City Archives, provided valuable information. Emily Hanna of State Records New South Wales helped with information concerning Sydney. Emily Chapin, Collections Access Archivist at the Museum of the City of New York, assisted with the Henry Bergh correspondence. David Allen responded with kindness to inquiries on the archival holdings of the RSPCA. Jan Holmquist of the Massachusetts Society for the Prevention of Cruelty to Animals answered my questions, and Debra A. Gardella of the Massachusetts Superior Court, Fall River and James P. Perkoski, First Assistant Clerk-Magistrate, Attleboro District Court, generously assisted in tracking down the remaining information on the 1910–1911 case from Attleboro. Gina Y. Hodges, Administrator of the Atlantic City Municipal Court searched for files on the Kaplowitz case. Josh Jasper, Librarian and Archivist at the Rhode Island Jewish Historical Association was invaluable in finding information on the relationship between the Jewish community in Providence and the Jews of rural Massachusetts and identifying Rabbi Katznelson. Kevin Luy of the Colorado State Archives provided the records of the Wolf Heller case. Vaughan Black and Sheila Wildman shared insights into the history of law and animals in Halifax.

This project started in the files of the Alex Dworkin Canadian Jewish Archives in Montreal. Janice Rosen and Hélène Lavallée are an inspiration and to them this is yet another expression of my gratitude. My sincere gratitude to Professor Ira Robinson and the Concordia Institute for Canadian Jewish Studies for their support in publishing this book. The opinions expressed herein are my own.

As always, this is for Kathryn.

Introduction: Jews, *Shechita,* and the Law

The Historical and Linguistic Stakes

In the spring of 1913, Abraham Levitt, rabbi of the small Jewish community in Halifax, Nova Scotia was charged on the information of an inspector of the Society for the Prevention of Cruelty, with inflicting ill-treatment on, and causing unnecessary suffering to, a heifer. According to the local newspaper coverage of the case, this was "the first time in American history that a rabbi has been haled before the courts in a case of this kind."[1]

After a hearing lasting several days, Levitt was convicted and fined six dollars plus seven dollars and sixty-five cents in costs. For the first time in the history of the American continent a Jewish rabbi had been fined for killing an animal according to the Jewish belief, namely by cutting its throat without prior stunning.[2] The historian of Montreal and Canadian Jewry, David Rome, later wrote that, "because Halifax is so distant from the centres of Canadian Jewry a very important anti-Semitic case in that city, classic in its form, received very little attention and has since been virtually lost to memory."[3]

This book remedies the gap identified by Rome by returning the Halifax case to our collective legal and historical memory. It interrogates the often-competing discourses of the anti-cruelty movement in the early twentieth century, narrow positivistic renderings of legal texts, and ideas of religious

1 "The Halifax Hebrew Rabbi Is Before Stipendiary Magistrate's Court," *Halifax Herald,* April 2, 1903.

2 "Jewish Rabbi Fined by the Stipendiary," *Halifax Herald,* April 22, 1913.

3 *The Jewish Biography of Henri Bourassa,* 94.

freedom and practice that circulated around, and informed the context of, the prosecution of Rabbi Levitt. It is necessary to place events in Halifax in 1913 in broader social, political, and socio-legal historical contexts that surrounded local, national, and international debates about animal welfare and about the place of growing Jewish communities in North America. At first blush, the singular prosecution of Abraham Levitt, apparently the only recorded instance of an animal cruelty case and conviction of a Jew engaged in the religiously ordained killing of an animal in North American legal history, might seem to be a minor historical event at the geographical and demographic margins of Canada, North America, and the British Empire. This study highlights the ways in which this singular instance of legal prosecution embodied matters of historical, social, and cultural significance that went far beyond the fate of Abraham Levitt and the small Jewish community of Halifax. It brings to light other instances of attempts to criminalize Jewish religious slaughter in the common-law world, demonstrating how and why the singular instance of the Halifax case was, contrary to coverage at the time, not at all singular. Instead it fits into a pattern of actions by animal welfare groups targeting the Jewish method of slaughtering animals for food over a significant period of time and in a variety of jurisdictions. I seek to recover for historical memory the vital role of these anti-*shechita* prosecutions in the legal and social history of Jewry in the different places they occurred.

By bringing to light these other attempts to criminalize and abolish *shechita*, the process of interrogating the hermeneutic and narrative similarities that informed each of these now somewhat obscured legal instances, in which observant Jews were prosecuted for practices central to their religious identity in the name of humane values, can begin. The prosecution of Abraham Levitt in Halifax on the eve of the First World War was not as singular or isolated as David Rome believed. An important goal of this study is to begin to trace a socio-legal history of attempts in the English-speaking world to criminalize the practice of the Jewish mode of killing animals in the name of the competing value of animal welfare and to interrogate the hows and whys of legal processes that contrast in a negative and harmful way, humane values and Jewish religious observance. Each instance highlights the common conflicts that occurred in the legal regulatory realm as a public good, animal welfare, encountered another public good, the need to recognize and accommodate religious practice. While each clearly occurred at different times and in distinct geographical spaces, neither Jewish communities nor animal welfare groups acted in complete isolation from their fellows, nor did they always operate in ignorance of the real legal history of attempts to criminalize the Jewish method of slaughter. Networks of knowledge, shared values and history, and group solidarity operated on both sides of these struggles.

The Halifax case of 1913 underscores the ways in which discourses of solidarity among Canadian Jewish communities were at play, as were national and transnational practices and networks of knowledge relating to the anti-cruelty movement in Canada, the United States, and the United Kingdom. The Halifax legal action must be understood not just in relation to the specificities of the case itself, but in these contexts and discourses that went well beyond a single conviction of a solitary Jew in a small Jewish community at the geographic and social fringes of the Canadian federation, the British Empire, and Canadian Jewry. The patterns that emerge in these cases will elucidate the fashion in which Jewish communities have been forced to struggle for recognition and equality in the midst of social, cultural, political (and legal) contexts that invoked a fundamental distinction between the normal and Christian human spirit and a Jewish world tainted by an equally Christian imaginary around the central trope of blood.[4]

Societies for the Prevention of Cruelty to Animals or Humane Societies did not hesitate to claim that their sole aim was the protection of "God's dumb animals" and that they acted from this moral and theological imperative. Jewish communities defended *shechita* in terms of their obligations as observant practitioners of their faith. The anti-*shechita* prosecutions bring to the light the difficult, if not impossible, task involved in the historical articulation of a broadly acceptable and accepted social, political, moral, and juridical position in cases where each claimant invokes holy writ in demanding the legal recognition and embodiment of its claims. When Abraham Levitt killed the heifer in a slaughterhouse in Halifax in the spring of 1913, he did so as part of his duties as *shochet*, an individual trained in the killing of animals according to the methods prescribed by Jewish law. For observant Jews, the shochet "is more like a High Priest, for he must be a god-like man, a skillful [*sic*] man in his profession; a learned man, who chooses this humble and dangerous calling, because he believes it his mission to so serve not only his generation; but the generations to come."[5]

---

4 Anthony Bale and David Feldman, eds., *Blood: Reflections on What Unites and Divides Us* (Oxford: Shire, 2016).

5 Friedman Family / Friedman, Aaron / Aaron's Life as Shochet / n.d. Friedman Family Papers; P-948; Box 1, Folder 1, American Jewish Historical Society, New York and Boston.

The cow had been dispatched according to the Jewish method of killing, *shechita*. While it is still common to deploy the terms "ritual killing" or "ritual slaughter" to describe shechita, I adopt the less morally fraught, and historically problematic, "Jewish method" or "Jewish mode" formulations. The slippage, and the linkage made, between accusations of Jewish ritual murder of Christian children in the blood libel, and a signifying chain of Jewish brutality and blood lust associated with shechita, has been present in anti-Judaic discourse and legal practice throughout the nineteenth and twentieth centuries.[6] Fantasies about Jews engaging in ritual murder were linked to the meanings associated with the reality of ritual slaughter at the turn of the century. Interlocking discourses of murder and ritual slaughter are present in the world of literary representation as well as in the rhetoric of the blood libel accusations of the time.[7]

Anthony Julius, in his monumental study of English antisemitism, examines the semantic and rhetorical elisions and connections between the use of terms such as ritual slaughter and the ritual murder accusation as antisemites joined animal welfare and anti-cruelty debates.[8] The two competing principles at play in the controversies over shechita, a practice performed as a result of religious obligation, and an opposing deontological assertion about the welfare of animals, call for carefully elaborated ethical normative engagements between the parties. However, the signifier "ritual slaughter" carries with it a clear message, especially when paired or contrasted with ideals of the humane, as Brian Klug has argued, of a contrast between a civilized and humane Us and a savage and merciless Other, the Jew.[9] Throughout modern British and other debates about the relationship between shechita and claims of animal suffering during slaughter, phrases such as "Jewish mode of slaughter," "the Jewish method," "the Jewish mode," or "Jewish rules" have been used interchangeably in order to avoid the potential slippage into pernicious and dangerous antisemitic tropes. Because antisemitism is one of the organizing discourses that informed debates in the cases of attempted criminalization of Jewish practices, I adopt Julius' preferred formulations when discussing shechita.

6 Richard Weisberg, ed., *A Thousand Years of Infamy: The History of the Blood Libel*, 28 Law and Literature (2016).

7 Sander L. Gilman, "Kafka Wept," *Modernism/Modernity* 1, no. 1 (1994):18.

8 *Trials of the Diaspora: A History of Anti-Semitism in England* (Oxford: Oxford University Press, 2010), 341–342.

9 "The animal welfare lobby is wrong," *Guardian*, June 11, 2003.

Shechita, Religious Obligation, and Animal Welfare

The approved mode of killing animals so that they might be consumed fits more broadly into the complex and multifaceted set of Jewish dietary laws known as kashrut. For the observant Jew, the dietary laws form an essential part of religious experience and being, although it is worth noting that kashrut does not compel individual Jews to consume meat. It simply provides that meat that is to be consumed must come from animals slaughtered according to the rules of shechita. The laws of kashrut are understood and practiced as religious and ethical obligations that go to the heart of Jewish existence. "To acknowledge God in all one's ways is a short expression of the fundamental conception of Judaism that divine worship is not confined to prayer or to religious service in the House of God but is inherent in every facet of life, even in the matter of sustenance."[10]

In shechita, the animal is slaughtered according to the prescriptions of the Torah and the Talmud.[11] In Jewish law, the animal must be sound, with no indication of illness or injury. Because the use of a bolt, electric shock, or other method, to stun the animal into unconsciousness, before killing, would by necessity cause physiological change, and injury, the Jewish method requires that the animal be killed without being stunned.[12] One swift cut with an extremely sharp knife, created specifically for this purpose, and itself subject to strict regulations (a *hallaf* or *sakin*), in the hands of a trained shochet "causes no pain, and takes a fraction of second."[13]

The case against the shochet in Halifax was formed within the semiotics the aim of which was protecting animals from unnecessary suffering. The prosecution charged that Rabbi Levitt, in pursuing the Jewish method, had caused cruelty to be inflicted on the animal. The evidence led by the authorities

10 I. Grunfeld, *The Religious and Moral Basis of the Jewish Dietary Laws* (London: National Council of Shechita Boards, 1966), 1.

11 The divine injunction informing the Jewish method of slaughter comes among other sources from the text of Deuteronomy XII, 21 and from the *Babylonian Talmud, Tractate Hullin*, and codifications by Maimonides and *Karo* and other sources. Bernard Homa, *Shehita* (London: Soncino Press/Board of Deputies of British Jews, 1967).

12 S.I. Levin and Edward A. Boyden, *The Kosher Code of the Orthodox Jew* (Minneapolis: University of Minnesota Press, 1940); C.G. Montefiore, "Dr. Wiener on the Dietary Laws," *Jewish Quarterly Review* 8, no. 3 (1896); and M. Hyamson, "Another Word on Jewish Dietary Laws," *Jewish Quarterly Review* 9, no. 2 (1897).

13 Michael L. and Eli Munk, eds., *Shechita: Part II of "Edut Ne'emana: Religious and Historical Research on the Jewish Method of Slaughter* (Brooklyn: Gur Aryeh, 1976), 11.

against Levitt alleged that the heifer issued loud noises as it thrashed about, evidencing almost in and of itself, the suffering and unnecessary pain inflicted by the shochet. The rabbi's defense, at the first stage at least, engaged with the animal cruelty allegation in terms that have informed the Jewish defense of shechita throughout the nineteenth, twentieth, and twenty-first centuries. For Levitt and his attorneys, as well as the medical experts who testified in his behalf, the Jewish method, with its swift single cut, was much less cruel than the common slaughterhouse method of animal stunning. This defense of the Jewish method of killing confronted allegations of animal cruelty directly and on their own terms, invoking modern medical and physiological science. Defenders of the Jewish method attacked the not always subtle assertions by opponents of shechita that the practice was a per se violation of a moral code demanding human kindness to animals. In a core text written to rebut animal protectionist allegations about the inherent and immoral cruelty of shechita, leading rabbi and shochet Aaron Z. Friedman offered both a physiological justification for the Jewish method and a theological/ethical explanation for the preferability of shechita. "And from the very beginning, when God created His creatures and ordained that animals should serve for human food, He commanded the Israelites to alleviate the pains and sufferings of animals by killing them in this kind way of Shechitah [*sic*]."[14]

Debates about the Jewish mode of slaughter embodied in the Halifax case gave voice to the multiple discourses that informed discussions and polemics about shechita throughout the nineteenth and twentieth centuries and that continue to this day. There were (and are) arguments about cruelty and ethical duties to animals, as well as unsubtle claims of the superiority of Christian morality over crude and primitive Jewish practice. The Halifax case, perhaps long forgotten or marginalized, offers useful insights into debates about animal cruelty and early manifestations of discourse about the meaning and content of ideals of religious freedom at the edges of Canada, North America, and the British Empire.

The Halifax Jewish Community

Writing in the early 1980s, Terrence Murphy described the state of Jewish history in Atlantic Canada as "still in its infancy."[15] While the stories of the Jewish

14 Laemlein Buttenwieser, trans., *Tub Taam or a Vindication of the Jewish Mode of Slaughtering Animals for Food Called Shechitah* (2nd ed) (New York: Bloch Publishing, 1904), 25.

15 "The Religious History of Atlantic Canada: The State of the Art," *Acadiensis* 15, no. 1 (1985): 173.

populations of the Atlantic provinces, and of Halifax, have been presented in subsequent studies, this history has been recounted largely as part of the broader Canadian or North American Jewish narrative,[16] or as an introduction to sociological or political accounts of the state of contemporary Jewish communities, particularly those situated outside the major population centers.[17] While this is not the place to offer a more complete history of the Halifax Jewish community, it is necessary to set out some context for the events surrounding Rabbi Levitt's conviction for animal cruelty in the spring and summer of 1913.

The historical accounts of Halifax Jewry that do exist offer a common narrative. A significant presence of Jewish merchants in Halifax can be found from the city's founding in 1749 as a key port and garrison during the British war with the French.[18] The period of the War of Independence saw a further influx of Jewish individuals from Rhode Island and other American colonies.[19] There was a strong enough presence of Jews in Halifax for the purchase and setting aside of land in the city's cemetery for a Jewish burial area, although there is no evidence that it was ever used for this purpose. Nor is there any documentation to support the existence of a Jewish burial society, a *Chevra Kaddisha*, a key historical indicator of an active Jewish community.[20] Over the next century the identifiable Jewish presence in Halifax dissipated and what was left was a diminishing number of individuals with "Jewish" surnames scattered throughout Nova Scotia, but little or no indication of a separate, identifiably Jewish community.

The influx of European Jews to North America in the 1880s and 1890s saw a renewed presence in Halifax. As the major port in Atlantic Canada, the city was a point of disembarkation for thousands of immigrants to North

---

16 Gerald Tulchinsky, *Taking Root: The Origins of the Canadian Jewish Community* (Hanover, NH: University Press of New England, 1993), 82 et seq.; Sheldon J. Godfrey and Judith C. Godfrey, *Search Out the Land: The Jews and the Growth of Equality in British Colonial America, 1740–1867* (Montreal: McGill-Queen's University Press, 1995), 73–81.

17 A. Ronald Gillis and Paul C. Whitehead, "The Halifax Jews: A Community within a Community," in *Immigrant Groups*, ed. Jean Leonard Elliott (Scarborough ON: Prentice Hall, 1971), 84–94; Sheva Medjuck, *Jews of Atlantic Canada* (St. John's, NL: Breakwater, 1986); Shari L. Brotman and James L. Torczyner, *The Changing Portrait of Jewish Life in Halifax* (Halifax, NS: McGill Consortium for Ethnicity and Strategic Social Planning, 1997).

18 Jacob Rader Marcus, *Early American Jewry: The Jews of New York, New England and Canada, 1649–1794* (Philadelphia: Jewish Publication Society of America, 1961), 200.

19 Tulchinsky, *Taking Root*, 83.

20 Godfrey and Godfrey, *Search Out the Land*, 77, 79; David Wachtel, "Jewish Burial Societies: The Origins and Development of the *Hevra Kaddisha*," *Jurist* 59 (1999).

America and has been described as "Canada's Ellis Island."[21] While many of these immigrants moved on to other parts of the country or south to the United States, some stayed and the Jewish community of Halifax was reborn. The 1881 Canadian census reveals a total of fifteen identified Jews, born in Germany, the United States, and Canada.[22] By 1891, while the numbers had barely changed, the place of birth of the individuals or their parents clearly indicates an increasing presence of Russian and Polish Jews in Halifax.[23] The growth in numbers at the time of the next census in 1901 is significant throughout the province and specifically in Halifax. In 1881 there were thirty-two Jews in the entire province of Nova Scotia and by 1901, the number had risen to 449.[24] The newly arrived Jews came from central and eastern Europe and occupied roles across the middle and lower middle class sections of the economy, working as tailors, traders, merchants, clothiers, jewelers, peddlers, dressmakers, junk dealers, and included an optician.[25] By 1911, two years before the shechita prosecution, the number of Nova Scotia Jews had reached 1,360.[26] While the population of the province grew by seven point thirty-five per cent in the decade 1901 to 1911, the still small Jewish population increased by over 200 per cent.[27] Tulchinsky points out that "these statistics reveal that the Jewish population of Canada was spreading out to smaller urban places in response to expanding economic opportunity."[28]

The Halifax Jewish community, now united as a congregation, but still without a synagogue, adopted its name in honor of the European philanthropist Baron de Hirsch. It played an active role in rendering assistance to newly arrived immigrants from Europe, those who wished to settle, and those who arrived with the intention of moving on.[29] In 1890, the secretary of the Jewish community in Halifax entered into contact with the first non-synagogal Jewish

---

21 Stuart Rosenberg, *The Canadian Jewish Community*, Vol.1 (Toronto: McClelland and Stewart, 1970), 108.

22 Glen Eker and Deborah Pekilis, *Index of Jews Resident in New Brunswick, Nova Scotia and Prince Edward Island According to the 1861 to 1901 Censuses of Canada* (Toronto: Ontario Genealogical Society, 2004), 9.

23 Ibid., 16–17.

24 Louis Rosenberg, *Canada's Jews: A Social and Economic Study of Jews in Canada in the 1930s*, ed. Morton Weinfeld (Montreal: McGill-Queen's University Press, 1993), 20.

25 Eker and Pekilis, *Index of Jews*, 36–39.

26 Rosenberg, *Canada's Jews*, 20.

27 Ibid., 12.

28 *Taking Root*, 159.

29 *Baron De Hirsch Congregation 1890 to 1990: 100th Anniversary Commemorative Book* (Halifax, NS: Beth Israel Synagogue, 1990).

organization in Canada, the Young Men's Hebrew Benevolent Society in Montreal, to seek assistance in extending relief to disembarking immigrants.[30] Jewish Haligonians still worshipped in a number of temporary facilities. In these early days of communal life, religious ceremonies were held in individual homes. As the number of Jews grew, services were moved to premises made available by fraternal organizations such as the Masons and Oddfellows. In the mid-1890s, the seventeen Jewish families in Halifax, with the assistance of "Christian friends" purchased and refurbished the former Free Baptist Church on Starr Street that was dedicated as the Baron de Hirsch synagogue in February 1895.[31] Throughout this time, relations with the dominant Christian communities of Halifax, particularly the various Protestant denominations, appear to have been warm. Leading Protestants assisted the community by providing facilities for worship and subscribed to the fundraising drive for the new synagogue. The local immigration agent in Halifax, Edward Cleary, himself a Free Baptist minister, was described as "very considerate" to newly arrived Jewish immigrants. The rejection rate for Jews seeking permission to land in Halifax was the lowest in Canada.[32]

This is not to suggest that the history of Jewish life in Halifax was or is devoid of antisemitism. The history of the province of Nova Scotia is underscored by the divisions between Roman Catholicism and various Protestant denominations, including the Anglican and Presbyterian elites. If intra-Christian religious disputes and an ultimate modus vivendi grounded in institutional structures recognizing the splits are at the heart of the history of Nova Scotia and Halifax, one could hardly expect an absence of Christian antisemitism. The prosecution of Rabbi Levitt cannot be understood without taking into account the grossly anti-Jewish motifs at play in the case and abroad in the country more generally.[33] At the same time, the history of the Halifax Jewish community in the 1890s and into the first years of the twentieth century was characterized by a significant amount of collaboration, cooperation, and good will, with and from significant parts of the Christian communities of the city.

---

30 Simon Belkin, *Through Narrow Gates: A Review of Jewish Immigration, Colonization and Immigrant Aid Work in Canada (1840–1940)* (Montreal: Canadian Jewish Congress/Jewish Colonization Association, 1966), 38–39, 88.

31 *Baron de Hirsch Congregation*, 25.

32 Jan Goeb, *History of the Jews in the Maritimes*, Undated typescript, Jewish Historical Society Fonds, MG 20-Vol. 561-7, Nova Scotia Archives, Halifax, 11.

33 Ira Robinson, *A History of Antisemitism in Canada* (Waterloo, ON: Wilfrid Laurier University Press, 2015), 54.

Victorian Reform, Protestant Ethics, and the Society for the Prevention of Cruelty in Halifax

Noted historian of Nova Scotia and Halifax, Judith Fingard, in writing about the Nova Scotia Society for the Prevention of Cruelty (SPC), observed that "the anti-cruelty movement, of which it was a part, has been virtually ignored by Canadian historians."[34] While I cannot remedy the situation and present an exhaustive historical account of the SPC or of the animal welfare movement more broadly, it is important to note the existence of contemporary national and international contexts of debates over animal cruelty and to underline the fact that the Nova Scotia SPC must be understood in terms that place it within the overall social and legal history of animal welfare organizations, as they originated in the United Kingdom,[35] and grew in Canada,[36] and the United States.[37] In the 1913 prosecution of Rabbi Levitt, the broader national and international contexts would come to play significant and under-explored roles.

As with anti-cruelty movements elsewhere, the Nova Scotia SPC (and SPCA as its nomenclature and functions changed) in its foundation and activities at the time of the Levitt case fits into the broader context of Victorian and Edwardian, or in the United States historical taxonomy, Gilded Age, social reform movements. James Turner demonstrates how the emergence of the SPCA in England ran in parallel with a period of fervent evangelical agitation.[38] Janet Davis traces the intellectual and theological origins of concern over animal welfare to periods of increased religiosity during the Second Great Awakening in the United States and the continuing influence of a particular Protestant ideology.

> ... this era of evangelical revivalism and social reform was essential to the birth of an organized animal welfare movement after the Civil War. The Second Great Awakening put a native-born Protestant stamp on

---

34 Judith Fingard, *The Dark Side of Life in Victorian Halifax* (Porters Lake, NS: Pottersfield Press, 1989), 171.

35 Antony Brown, *Who Cares for Animals? 150 Years of the RSPCA* (London: Heinemann, 1974).

36 Beatrice Johnston, *For Those Who Cannot Speak: A History of the Canadian Society for the Prevention of Cruelty to Animals, 1869–1969* (Chomedy, PQ: Dev-Sco, 1970); *The Canadian Society for the Prevention of Cruelty to Animals* (Montreal: CSPCA, 1886).

37 Diane L. Beers, *For the Prevention of Cruelty: The History and Legacy of Animal Rights Activism in the United States* (Athens: Ohio State University Press/Swallow Press, 2006).

38 James Turner, *Reckoning with the Beast: Animals, Pain, and Humanity in the Victorian Mind* (Baltimore: Johns Hopkins University Press, 1980), 20–22.

> the universalizing mission of animal kindness. The Awakening stressed the potential for human perfectibility on this earth, rather than passive acceptance of God's will. Religious revivalism highlighted the power of free moral agency to combat social injustice.[39]

Grounded in a deeply and overtly Christian moral vision, these movements sought greater protections and rights for women and children, as well as animals. "They made "cruelty" into a social problem and wedded sentimentalism to liberal rights discourse to expand public responsibility for animals and children."[40]

In Halifax and elsewhere, they sought to inculcate a proper moral vision and practice, and to improve general welfare with an emphasis on the links between a proper education and the embodiment of moral values.[41] Social and fraternal organizations not only forged the basis for a rising middle class identity in Halifax, but concretized denominational solidarity. They later served as the structural bases upon which the purely private associational form took on public functions of welfare and reform, especially in terms of temperance and the social emphasis on the moral educational value of work.[42] These organizations battled against the worst consequences of moral failure, often infused with issues of alcohol abuse, in their campaigns against rampant criminality and especially prostitution.[43] They combatted the sources of cruelty and neglect in their struggles for temperance.

> In Canada the rapid industrialization and urbanization of the Laurier era created or threw into sharp relief a host of social ills. Red light districts abounded in the towns and cities, alcoholism increased sharply, the exploitation of workers became blatant and the failure of traditional institutions to provide security for the less fortunate was increasingly manifest. Rural residents were alarmed not only by the moral and social problems of the cities and towns but also by the depopulation of their own

---

39 Janet M. Davis, *The Gospel of Kindness: Animal Welfare and the Making of Modern America* (Oxford: Oxford University Press, 2016), 27–28.

40 Susan J. Pearson, *The Rights of the Defenseless: Protecting Animals and Children in Gilded Age America* (Chicago: University of Chicago Press, 2011), 19.

41 Robert Nicholas Bérard, "Moral Education in Nova Scotia," *Acadiensis* 14, no. 1 (1984).

42 David A. Sutherland, "Voluntary Societies and the Process of Middle-class Formation in Early Victorian Nova Scotia," *Journal of the Canadian Historical Association* 5, no. 1 (1994).

43 Fingard, *The Dark Side of Life in Victorian Halifax.*

> communities. Nova Scotians, who were noted for their strong church allegiance, tended to look to the clergy for leadership in solving their problems. The latter proposed as a general solution implementation of the social gospel— a fundamental reform of society on the basis of Christian principles.[44]

The industrialization and commercial growth in Canada that had presented immigrant Jews with opportunities to establish and to build a community in Halifax had created a city of conflict, criminality, and social ills. It is hardly surprising, given the history and sociology of Nova Scotia as fundamentally Christian, divided between Protestant denominations and the Roman Catholic Church, that the social reform movements would not only be influenced by religious tenets, but that the institutional structures put into place would reflect the faith-based distinctions abroad more generally in the city.[45] Given the significant presence of African Canadians in Halifax since the days of the United Empire Loyalists, a distinct set of racially segregationist structures and norms also operated in the city and its social welfare apparatus.

Renée Lafferty emphasizes the core values of denominational divisions in the ideology and culture, as well as the organizing principles at work in child welfare. "Denominational partitions dismissed by some as unnecessary (or even damaging) were considered by others to be vital to ensuring a child's future—this future being thoroughly entwined with the specific culture into which the child was born."[46] Social welfare in Halifax from Victorian times onwards was infused with Christian values and embedded in denominational structures. Significantly, Judith Fingard described the anti-cruelty movement in Halifax, embodied in the SPC, as at one and the same time the "most active interventionist society in late nineteenth-century Halifax" and as "more secular in its orientation than other voluntary organizations."[47]

---

44 E.R. Forbes, "Prohibition and the Social Gospel in Nova Scotia," *Acadiensis* 1, no.1 (1971): 14.

45 David A. Sutherland, "Race Relations in Halifax, Nova Scotia, During the Mid-Victorian Quest for Reform," *Journal of the Canadian Historical Association* 7, no. 1 (1996); Judith Fingard, "Race and Respectability in Victorian Halifax," *Journal of Imperial and Commonwealth History* 20, no. 2 (1992).

46 *The Guardianship of Best Interests: Institutional Care for the Children of the Poor in Halifax, 1850–1960* (Montreal: McGill-Queen's University Press, 2013), 21.

47 *The Dark Side of Life in Victorian Halifax*, 171; "The Prevention of Cruelty, Marriage Breakdown and the Rights of Wives in Nova Scotia, 1880–1900," *Acadiensis* 22, no. 2 (1993). In 1880, the Nova Scotia legislature changed the law to extend the powers of the SPCA especially in child protective matters in An Act to amend the Act to incorporate the

The first national anti-animal cruelty provisions in Canada were enacted in 1869.[48] In the same year, the Canadian Society for the Prevention of Cruelty to Animals was incorporated in Quebec.[49] Among the founders and early leaders of the Society in Montreal were the Joseph brothers, Jacob Henry and Jesse. Members of an established Quebec Jewish family, they were prominent businessmen, who took active roles in various commercial enterprises, and in public service and charity work. Jacob was identified with the Montreal General Hospital and Jesse was for many years a prominent and active member of Shearith Israel, the Spanish and Portuguese synagogue, the oldest in Canada.[50] Joining the Josephs among the earliest public supporters of the Society and its work for animal welfare was the Rev. Meldola de Sola.[51] Meldola de Sola was the son of the Rev. Dr. Abraham de Sola, longtime rabbi of the Spanish and Portuguese Synagogue. Meldola succeeded his father in 1882 and became the first Canadian-born rabbi to serve a congregation in that country.[52] At a time when many Protestant ministers failed to offer their public backing for the Society, de Sola was one of the few religious leaders in Montreal to deliver a sermon on kindness to animals as a core ethical value and was staunch in his support of the Society and its goals.[53] In its early days in Montreal, the Society's leadership featured some of the most prominent and influential members of the city's Jewish community. In its beginnings, it did not proclaim the overtly Christian evangelical ideology of the English, or other Canadian Societies, but instead it was more broadly humanist in its orientation, although from its foundation it was dominated by the English-speaking Protestant social and commercial elite of the city.

In Nova Scotia, the SPCA was incorporated eight years later in 1877. The SPCA itself makes much of the history of Nova Scotia as a groundbreaking jurisdiction in the field of the punishment of animal cruelty, but a more

---

Nova Scotia Society for the Prevention of Cruelty to Animals, 43 Vict. c. 68 and in 1909 the broader remit of the SPCA was recognized as it was renamed in section 1, "The Nova Scotia Society for the Prevention of Cruelty," An Act to amend chapter 86 of the Acts of 1877, entitled "An Act to incorporate the Nova Society for the Prevention of Cruelty to Animals," 9 Edw. VII, c. 165.

48 An Act Respecting Cruelty to Animals, SC 1869, c. 27; 32–33 Vict., c. 27.

49 Johnston, *For Those Who Cannot Speak.*

50 Arthur Daniel Hart, ed., "Jacob Henry Joseph," in *The Jew in Canada* (Toronto: Jewish Publications, 1926), 330; "Jesse Joseph," ibid., 331.

51 Johnston, *For Those Who Cannot Speak,* 8.

52 Hart, *The Jew in Canada,* 87.

53 Johnston, *For Those Who Cannot Speak,* 8.

careful examination of the record puts a somewhat different spin on matters. The modern day claim by the SPCA is that "the Acts of Nova Scotia made provisions for public whipping to be the punishment for persons convicted of cruelty to animals."[54]

The assertion becomes less significant when the legislative context and the 1824 statute lauded by the Society itself are examined more carefully. The 1824 statute, An Act to punish Persons guilty of maliciously killing or maiming Cattle,[55] replaced the similarly titled Act to prevent the malicious killing or maiming of cattle, from 1768.[56] A plain reading of the laws indicates that the provisions are, unsurprisingly given the historical and legal contexts in which they were passed, not aimed at animal welfare per se, or at the prevention of cruelty as a moral imperative. The statutes granted relief "by action of trespass or upon the case" to the animal's owner. The 1768 law gave the owner the right to recover treble damages against anyone who "maliciously, unlawfully and willingly" killed or maimed an animal not belonging to him. The 1824 law, that introduced the penalty of whipping, repeated these provisions of the earlier Act, including the use of the "maliciously, unlawfully and willingly," but not "cruelly." The penalty of whipping was an alternative to the usual punishment of imprisonment and was solely within the discretion of the Court.[57] Imprisonment or whipping was not permitted if the owner had proceeded by way of damages, and conversely, the infliction of the punishment barred any subsequent action for monetary recompense.

Whipping was not, as the SPCA narrative would have it, the punishment for cruelty to animals. It was a punishment available in some cases, largely as a discretionary alternative. The action that gave rise to the punishment was not characterized as "cruelty" but required malice and unlawfulness. The Nova Scotia statutes were little more than treble damages provisions within the normal field of tort law relating to trespass to chattels and similar and related general actions in civil liability. The actions targeted by the legislation were those of the legal stranger, a third party who "maliciously" interfered with the property rights of "the party aggrieved." There is no mention of cruelty because cruelty was completely irrelevant to the legal issues of title to the animal. Killing, maiming, wounding, or otherwise hurting an animal could be

---

54 "History of the Nova Scotia SPCA," http;//spcans.ca/old_joomla_site/about-us/history.html.

55 Nova Scotia Statutes at Large, 4 & 5 Geo. IV, c. IV.

56 Nova Scotia Statutes at large, 8 Geo. III.

57 Nova Scotia Statutes at Large, 4 & 5 Geo. IV, c. IV, II.

done, especially in so far as killing was concerned, in a humane way, but would still constitute an unlawful interference with the rights of the owner of the animal. An owner of the animal could, within this statute, kill, maim, wound, or hurt his own animal with impunity. As Lesli Bisgould points out, this idea of the animal as chattel has not only informed much of the legal debate over cruelty legislation, but it is clearly one that is situated at an entirely different normative level than a focus on the suffering of the animal as the core harm to be dealt with in a legal framework.[58]

Notwithstanding this attempt to rewrite the legal history of its own actions and the basis of the prevention of cruelty to animals in Nova Scotia law, the SPCA was among the first organizations of its type in Canada and one of the most active anti-cruelty groups in the country. The SPCA was incorporated by provincial statute in 1877 and the prevention of cruelty to animals and the "conviction of persons offending against the same" were its key organizational objectives.[59] From its origins, the SPCA was permitted to call upon the police to assist in its tasks and to have special officers appointed as agents of the Society to engage in enforcement actions.[60] One of the most intriguing aspects of the Halifax shechita case can be found in this official legislative and bureaucratic melding of private and public functions that characterizes the existence of the SPCA. An incorporated body is invested with the protection of a legislatively identified public good, preventing animal cruelty. [61]

This change in the nature of the legislation from the early statutes in Nova Scotia is at the heart of debates over the criminalization of the Jewish mode of slaughter and those instances of prosecutions examined here. Anti-cruelty legislation identifies the protection of animals, and ensures their humane treatment, because these values are incorporated into legislation as public goods, as matters of general interest and value in which the state justifiably interferes. In addition to the rhetorical and juridical transformation of animal cruelty into an issue of public morality, for the SPCA in Nova Scotia and its counterparts elsewhere, "as private organizations delegated police powers, they

---

58 *Animals and the Law* (Toronto: Irwin Law, 2011), 34 et seq.; Joyce Tischler, *The History of Animal Law, Part I (1972–1987)*, 1 Stanford Journal of Animal Law and Policy 1 (2008).

59 An Act to incorporate, "The Nova Scotia Society for the Prevention of Cruelty to Animals," Statutes of Nova Scotia, 40 Vict., c. 86, § 4.

60 Ibid., and An Act to amend the Act to incorporate the Nova Scotia Society for the Prevention of Cruelty to Animals, Statutes of Nova Scotia, 43 Vict., c. 68 (1880).

61 Angus Nurse, *Beyond the Property Debate: Animal Welfare as a Public Good*, 19 Contemporary Justice Review 174 (2016).

animal cruelty for practicing shechita.[92] In praising the courage and fearless approach of Fielding in entering a conviction against the shochet, attorney Murray stated that "the Jewish method of slaughter has been discussed a great deal in England and different parts of America but no cases have yet been brought under the statute."[93] In his closing argument before Judge Wallace, W. J. O'Hearn echoed these statements when he declared that "it was extraordinary that in Great Britain, from which we copy the very legislation involved, and which country we regard as a model in the administration of justice, that no prosecution of this nature has ever been instituted against Hebrews."[94]

For all participants and many key observers of the trials of Abraham Levitt, the Halifax shechita case was a unique and potentially dangerous precedent, or conversely an important step in the legal struggle against cruelty. From March to June 1913, the Halifax Jewish community was at the center of national and international attention, from Jewish communities and from the anti-cruelty movement. David Rome later wrote with some regret that the Halifax case had "been virtually lost to memory."[95] This brief discussion may serve in some small way to revive the legal history of shechita in Nova Scotia and perhaps place Maritime Jewry, however transitorily, in a more central position in Canadian and North American Jewish (legal) memory.

But, without seeking to diminish the importance of the Halifax case of 1913, a more complete socio-legal history of shechita in the English-speaking world compels a demystification of the uniqueness argument advanced by all sides in Nova Scotia. The history of anti-cruelty laws against the practices of the Jewish method of slaughter is a longer and more complex one than Murray or O'Hearn, or the journalists of the *Halifax Herald* knew. Instances of prosecutions against the practice of shechita, from the earliest days of Societies for the Prevention of Cruelty, can be found in the legal histories of the United Kingdom, the United States, and even as far afield in the empire as Australia. Sadly, it turns out, claims of Nova Scotian uniqueness are in need of further nuance and qualification.

---

92 "The Halifax Hebrew Rabbi is Before Stipendiary Magistrate's Court."

93 "A Campaign Against the Butchers."

94 "Conviction in Jewish Cruelty Case was Quashed."

95 *The Jewish Biography of Henri Bourassa*, 94; B.G. Sack, *Canadian Jews–Early in This Century* (Montreal: Canadian Jewish Congress National Archives, 1975), 82 et seq.

extended the arms of government because they helped create and enforce new laws."[62] When a public norm is protected by such legislative intervention, a public agency, police or a government department, are tasked with ensuring the enforcement of the law through preventing offending behavior, and if that fails, with ensuring the pursuit of offenders through the courts. In animal cruelty cases, however, the Nova Scotia legislature, like that in many other jurisdictions, delegated the prevention, enforcement, and prosecution powers to a private body, the SPCA. That private body had, and still has, the power to investigate, to seize property, to charge, and to bring cases before the courts. In other words, they are the animal police, and can and do invoke the authority and powers of the courts, the institutions of the public enforcement of laws, as a private body enforcing public morals.[63]

In this sense, the 1913 summons of the shochet was a strong embodiment of the move away from social welfare and rehabilitation organizations and programs, towards a greater institutional emphasis on the policing of the types of anti-social behavior that had been targeted by fraternal and other bodies in the pre-World War I era. Social welfare intervention and rehabilitation were soon replaced by an emphasis on law and order.[64] The structures of law and order in Halifax, from the police to the courts, continued to embody many of the class and denominational characteristics of earlier reform movements.[65] The SPCA was no exception.

In cases like the Halifax prosecution, the SPC (re-named following its increased jurisdiction in child welfare) acted as investigator and prosecutor, combining the functions of police and attorney general, always in the name of the public, legislatively enshrined value of preventing cruelty to animals. The SPC was a single-issue body. It is focused on the prevention of cruelty. That was its private, corporate raison d'être and its public mandate. Unlike politicians or other public officials, it took, in a manner completely consistent with its public / private functions, a view of the public good and normative values that was singular in its focus. Animal welfare and the prevention of

62 Pearson, *The Rights of the Defenseless*, 19.

63 Jerry L. Anderson, *The Origins and Efficacy of Private Enforcement of Animal Cruelty Law in Britain*, 17 Drake Journal of Agricultural Law 263 (2012); Gordon Hughes and Claire Lawson, *RSPCA and the criminology of social* control, 55 Crime, Law and Social Change 375 (2011).

64 Michael Boudreau, *City of Order: Crime and Society in Halifax, 1918–35* (Vancouver: University of British Columbia Press, 2012).

65 Greg Marquis, "The History of Policing in the Maritime Provinces," *Urban History Review* 19, no. 2 (1990).

cruelty, these were the aims of the investigative, charging, and prosecutorial functions of the SPC. Other public values, unless they formed part of a broader legislative scheme on the protection of animals, could not and would not enter into the SPC's calculations, or act as determining factors in its decisions to prosecute acts of cruelty. The ideas of religious freedom, of core communal self-identity among Halifax Jews, of commandments from God to kill animals in a particular fashion, none of these could or did form the frame of reference for the SPC, whose sole aim was the prevention of cruelty to animals, women, and children, God's helpless creatures. A state prosecutor might well have considered the public interest more broadly to take into account possibly competing normative concerns, and have engaged in a specific and calculated balancing assessment of animal welfare on the one hand, and religious freedom on the other, in deciding whether to prosecute Rabbi Levitt for the practice of shechita. An organization such as the SPC, acting under its legislative mandate and in accordance with its own internal normative frame focused solely on the fundamental imperative of protecting animals, could not imagine in 1913 that any competing values might be at play. The case presented a chance to prosecute an obvious instance of inflicting needless suffering on an animal. The SPC wished to use the prosecution of the shochet to begin a campaign of education about, and pursuit of, slaughtering practices throughout the province that did not use stunning before killing.[66]

66 "A Campaign Against Butchers," *Halifax Herald*, April 28, 1913.

CHAPTER 1

Criminalizing Shechita: The Halifax Prosecution of Abraham Levitt, 1913

Halifax Jewry in 1913

While small in absolute numbers, by 1913 the Halifax Jewish community was well-established. They constituted an organized, functioning religious community with its own synagogue and rabbi, to serve spiritual needs, and to act as shochet. Halifax was in some sense a case apart in broader Maritime Jewish life. For more isolated Jewish settlements, so-called "Circuit Riding Rabbis," young men with sufficient, if incomplete, education, acted as shochet and *mohel,* and officiated at weddings or burials, earning their living as peddlers. They met the fundamental religious needs of scattered Jews. But what the Halifax Jewish community shared with their more isolated brethren was a set of core beliefs and practices where "Orthodoxy was the rule, not the exception in the Maritimes."[1]

Since colonial times, North American Jews had struggled with the real conflicts between religious obligations in relation to dietary laws, and, especially in the earliest days of European settlement, the practical and often insurmountable difficulties they confronted in complying with their duties in a deeply Christian setting. Some were able to maintain their religious practice in the privacy of their own homes, while abjuring strict obedience in public and during interactions with their Christian fellow citizens.[2] They might still, for example, not consume pork even in the public domain, maintaining this as a form of Jewish practice. From the earliest days of a Jewish presence in the

1 Goeb, *History of the Jews in the Maritimes,* 8–9.

2 Jonathan D. Sarna, *American Judaism: A History* (New Haven, CT: Yale University Press, 2004), 20; Jeffrey S. Gurock, *Orthodox Jews in America* (Bloomington: Indiana University Press, 2009), 24.

American colonies, shochetim were appointed and approved by rabbis and they did practice shechita.[3]

Despite the presence of some qualified shochetim in North America, Jews often would not, or could not, fully comply with the rules of kashrut, because there was no kosher food available.[4] Even in cases where growing Jewish communities were able to find a shochet,[5] or employ one of the "Circuit Riding Rabbis," especially for High Holy Days, serious doubts were cast by some on the training and skills of such butchers, and on whether animals were being killed in full accord with religious requirements.[6] The problem even extended to arguments and doubts about the ability of rabbis in America to issue convincing and binding pronouncements on many elements of Jewish law (*Halacha*), included in which would be aspects of the laws of kashrut.[7] But the influx of tens of thousands of Orthodox Jewish immigrants brought not just an increased demand for kosher food, but a supply of properly trained rabbis and shochetim who were able to meet that demand for the proper observance of religious law.[8]

For the Orthodox Jews of Halifax in 1913, the rules of kashrut were central to their communal identity and to their religious practice. Rabbi Levitt's role as shochet would have been at the heart of the life of the small number of Halifax Jews. But this desire to live according to holy command would come into conflict with other self-proclaimed ethically mandated and divinely ordained ideals of humane treatment of animals. Shechita met the SPC and the Nova Scotia legal system.

The Trial

On March 19, 1913, Andrew Williamson, a young Scot employed as an inspector for the SPC, together with an independent veterinary surgeon, Dr. Philip Gough, attended Shortell's slaughterhouse. They saw Rabbi Levitt

3 N. Taylor Phillips, "Family History of the Reverend David Mendez Machado," *Publications of the American Jewish Historical Society* 2 (1894): 48, 51.

4 Howard B. Rock, *Haven of Liberty: New York Jews in the New World, 1654–1865* (New York: New York University Press, 2012), 50; Jeremiah J. Berman, "The Trend in Religious Observance in Mid-Nineteenth Century America," *Publications of the American Jewish Historical Society* 37 (1947).

5 Gurock, *Orthodox Jews in America*, 61.

6 Ibid., 52.

7 Moshe D. Sherman, "Struggle for Legitimacy: The Orthodox Rabbinate in Mid-Nineteenth Century America," *Jewish History* 10 no. 1 (1996).

8 Anne Polland and Daniel Soyer, *Emerging Metropolis: New York Jews in the Age of Immigration, 1840–1920* (New York: New York University Press, 2012), 127.

kill a heifer using the Jewish method. According to the inspector and the vet, the heifer took at least fourteen minutes to die. Rabbi Levitt was summonsed to appear before the stipendiary magistrate to answer charges of causing unnecessary suffering and pain to the animal.[9] On April Fool's Day 1913, the shochet of the Orthodox Jewish community of Halifax faced his accusers. The ancient practice of killing animals for human consumption according to the Jewish method would now be judged before the courts of the province of Nova Scotia, for what was believed to be "the first time in American history."[10]

At the time of the shechita prosecution, the Nova Scotian courts and legal practitioners were in the middle of a process of transformation and increasing professionalization.[11] This was the result of the complex political and other factors that informed the broader practices and policies underpinning and arising from the economic and social reforms of the era.[12] While they had their own particular professional and biographical characteristics, the profession achieved a state of equilibrium on broad issues, such as class and religion, and on race in so far as African Canadian lawyers were concerned. "It is not terribly surprising that Catholics and Protestants managed to accommodate each other in this regard, as they has [*sic*] a fairly good record of doing so, in Halifax at least."[13]

As far as the judiciary is concerned, Girard identifies a lengthy process following which political patronage in appointments was replaced by criteria of professional skill and competence, as the Nova Scotia judicial branch "establish(ed) a certain legitimacy in the early twentieth century."[14] The first Roman Catholic judge of a superior court was not appointed until 1873, and before 1900, only three Roman Catholics had sat on the Nova Scotia bench.[15] While

9 "The Halifax Hebrew Rabbi Is Before Stipendiary Magistrate's Court," *Halifax Herald,* April 2, 1903.

10 Ibid.

11 Barry Cahill, "Birth of a Lawyer: James McGregor Stewart and the Halifax Bar on the Eve of Great War," *Journal of the Royal Nova Scotia Historical Society* 3 (2000).

12 Philip Girard, *The Roots of a Professional Renaissance: Lawyers in Nova Scotia 1850–1910,* 20 Manitoba Law Journal 148 (1991).

13 Ibid., 176.

14 *The Supreme Court of Nova Scotia, Responsible Government, and the Quest of Legitimacy, 1850–1920,* 17 Dalhousie Law Journal 430 (1994), 456; Philip Girard, Jim Phillips, and Barry Cahill, eds., *The Supreme Court of Nova Scotia, 1754–2004: From Imperial Bastion to Provincial Oracle* (Toronto: University of Toronto Press/Osgoode Society for Canadian Legal History, 2004).

15 Clara Creco, "The Superior Court Judiciary of Nova Scotia, 1754–1900: A Collective Biography," in *Essays in the History of Canadian Law,* eds. Philip Girard and Jim Phillips (Toronto: University of Toronto Press/Osgoode Society for Canadian Legal History, 1990), 51.

the situation did improve somewhat in later years, especially as the decline of Anglican influence opened up space for other Protestant denominations, a de facto quota of Catholic seats remained in effect for many years.[16] The first Jewish judge in Nova Scotia was not appointed until 1967, the centenary of Canadian Confederation.[17]

Abraham Levitt was summonsed to appear before the stipendiary magistrate on charges of animal cruelty. These judicial officers were the first point of call for law-breakers in Halifax.[18] In other parts of the province, magistrates were often not even legally trained. In many cases, stipendiary magistrates did not receive their costs for the hearing unless a conviction was entered. This gave rise to significant dissatisfaction with the lower levels of the criminal justice system in the province, and most significantly to a perception that magistrates and police were branches of the same enforcement agency.[19] In Halifax, however, the situation was less problematic and more recognizable as embodying separation of powers principles. The law provided that the stipendiary magistrate for the City of Halifax had to be a barrister of the Supreme Court of Nova Scotia with no less than five years' experience and that he hold no other position nor practice his profession. This was a full time judicial appointment.[20] At the time of the shechita trial, the stipendiary magistrate in Halifax was G.H. Fielding, who came from a prominent Baptist family, and was the brother of W.S. Fielding, former Premier of Nova Scotia, and a federal cabinet minister and Privy Councilor.[21] Throughout his tenure, Fielding was known as a merciful jurist.[22] On the face of it, a competent, professional, and compassionate jurist would determine Rabbi Levitt's fate.

Robert H. Murray led the prosecution. In his day job, Murray was a crown prosecutor, but he was and is perhaps "specially well-known as the counsel for the Society for the Protection [*sic*] of Cruelty."[23] From a prominent

---

16 R. Blake Brown and Susan S. Jones, "A Collective Biography of the Supreme Court Judiciary in Nova Scotia, 1900–2000," in *The Supreme Court of Nova Scotia,* 212.

17 Ibid., 212–213. Nathan Green, a Jew, was appointed as a Provincial Magistrate in 1959, and became first Chief Judge of the Provincial Court in 1981, "Chief Judge Nathan Green," in Judge R. E. Kimball, *The Bench: The History of Nova Scotia's Provincial Courts* (Halifax: Province of Nova Scotia, 1989), 65.

18 Boudreau, *City of Order,* 53–54.

19 Kimball, *The Bench,* 2–6.

20 Laws relating to the city of Halifax, 1891, § 133.

21 "Fielding, William Stevens," in Shirley B. Elliott, ed., *The Legislative Assembly of Nova Scotia, 1758–1983: A Biographical Directory* (Halifax: Province of Nova Scotia, 1984), 70.

22 Boudreau, *City of Order,* 54.

23 *Halifax Herald,* "Wedding," April 16, 1915.

Nova Scotia Presbyterian family, his work for the SPC, as well as his job as a criminal prosecutor, brought him into contact with many of the social ills of late Victorian and early Edwardian Halifax, from problems of wayward girls to conditions in the local jail, and the prevalence of cruelty.[24] He was a prominent member of the Halifax Civic Reform League, the organization that led the fight in pre-World War I Halifax against prostitution.[25] But in the shechita case of 1913 he wore not his crown prosecutor's hat, but that of counsel for the SPC, tasked with wiping out cruelty to animals in the city's and the province's slaughterhouses.

W. J. O'Hearn, K.C., defended Levitt. One of the best lawyers in Halifax and a leading expert on Canadian criminal law, O'Hearn was a Roman Catholic. He later served as Attorney-General of Nova Scotia, and upon leaving politics, sat as a County Court judge.[26] The prosecutor and the magistrate in the shechita case's first stages were prominent Protestants and the lawyer defending the Jewish shochet was a member of the Roman Catholic religious minority. At the time of his trial, Rabbi Levitt was described as "an old man, with a long flowing beard."[27] Of "Russian" origin, Rabbi Levitt did not speak or understand English. Samuel Webber acted as translator.[28] Webber was a member of a large and prominent Halifax Jewish family. He had arrived in Halifax from Russia in 1899, with his parents, Copel and Tona, and his many siblings. Webber's father, Copel, is described as a vegetable merchant, and acted as "the first Kosher Butcher until 1911" in Halifax.[29] Shortly after the case, the Webbers and others would split from the main community and form their own synagogue, known as the Proctor Street or Webber Synagogue. The rift, largely between the English-speaking "German" Jews and the more recently arrived eastern European Yiddish-speaking members would be healed some years later.

The case proceeded along what can only be described as parallel normative lines. Murray for the SPC introduced the testimony of Williamson and Gough to establish that they had witnessed the prolonged and unnecessary death throes of the heifer. Cross-examined by O'Hearn, the SPC inspector stated that he "inferred that the animal suffered pain because it struggled, and he based that statement on observations. In his opinion, the more humane way

24 Boudreau, *City of Order*, 43, 66, 89.

25 Ibid., 135–136.

26 "O'Hearn, Walter Joseph Aloysius," in Elliott, *The Legislative Assembly of Nova Scotia*, 171.

27 "The Halifax Hebrew Rabbi."

28 "Hebrew Rabbi Again Before City Stipendiary," *Halifax Herald*, April 16, 1913.

29 "Morris Herman and Rosa Bella," in *Baron de Hirsch Congregation*, 87.

was to first stun an animal before cutting its throat."[30] While Williamson had established some credentials as an inspector of the SPC, and as someone who had been present at many instances of killing cattle in his native Scotland, his opinion as to the suffering of the animal in this instance was clearly a matter of inference and opinion, from a non-scientific perspective, and O'Hearn tried to make much of this.

Murray sought to introduce objective proof of the unnecessary suffering inflicted on the heifer by Levitt in this case through the scientific testimony of Dr. Gough. He repeated Williamson's view that stunning the animal before cutting its throat was "the most humane method of slaughter."[31] What he had witnessed at the slaughterhouse in March as the animal thrashed about was clear evidence of suffering and in particular its attempts to bellow were irrefutable proof that the nervous system was still intact. For the SPC, this was a case where the cruelty and suffering inflicted on the heifer was a direct, avoidable result of the practice of shechita. Had the animal been stunned before its throat was cut, a more humane killing would have occurred. The failure to stun was what led directly to the commission of the cruelty offense.

The defense proceeded with a two-pronged counterattack. O'Hearn attempted to establish in his cross-examination of Gough that he could not know that the animal had suffered after its throat was slit and that like Williamson he was offering mere opinion and inference. Gough retorted that his answer to the questions was based on his knowledge as a veterinary surgeon of the science of the central nervous systems of animals, combined with what he had observed as an expert on March 19 in the slaughterhouse. Stipendiary Magistrate Fielding intervened to assert "... it is a fact because he says he has experimented. It is scientific knowledge."[32] The second part of the defense strategy would relate more directly to the practice of shechita. O'Hearn argued that Rabbi Levitt had done nothing different in Shortell's slaughterhouse than shochetim had done for thousands of years. He had followed the rules and practices of Judaism. Combined with this invocation of religious liberty, O'Hearn tried to establish that the Jewish method of slaughter was humane and did not involve the infliction of unnecessary suffering on the animal.

This defense against animal cruelty charges had been and continues to be commonplace among supporters of shechita. The argument is that because the

---

30 "The Halifax Hebrew Rabbi."

31 Ibid.

32 Ibid.

Jewish method involves one single and swift cutting blow with the *hallaf* that severs the blood supply to the brain and the central nervous system, the animal is killed quickly and painlessly. What Williamson and Gough described as the heifer thrashing and suffering for fourteen minutes, was an automatic nervous reaction known to occur in many animals after death. It was no proof of pain or suffering, since pain and suffering required a conscious or functioning brain. In other words, this part of the defense of shechita addresses the core element of the statutory offense: that the animal had needlessly suffered. This is a more strictly positivistic and legalistic defense than one grounded in vaguely formulated and legally ambiguous religious freedom terms in post-Edwardian Nova Scotia. As far as the defense was concerned, the SPC could not establish its case because shechita could be scientifically proven to be at least as humane as other methods, including the use of stunning. There could be no needless or unnecessary suffering, and no cruelty, if the method of shechita were at least as painless and quick as any other method then in use. The magistrate could acquit the defendant based solely in this legal scientific argument without having to enter the potentially trickier and more problematic debates about religious liberty.

O'Hearn continued with this line of attack.[33] He pointed out as he opened the defense's case that the law did not prohibit the infliction of suffering per se on animals, a principle that would have made all killing problematic. O'Hearn continued to meld his approach with the two branches of his position. He insisted both that the Jewish method did not cause unnecessary suffering and further that none of the prosecution's evidence from Williamson and Gough in fact or in law established any cruelty. He added that the law relating to unnecessary suffering had to be understood in such a way as to recognize the Jewish method as necessary because of the religious obligations that underpinned the practice. Magistrate Fielding was clearly not impressed with this line of argument and intervened in O'Hearn's opening statement to ask "whether conscientious objection" was a legal defense.[34] After O'Hearn had finished repeating his arguments that the prosecution had not made out a colorable case and that religious practice and belief had to be taken into account in interpreting the nature and extent of the infliction of unnecessary suffering on an animal under the statute, Fielding laid out his own jurisprudential position, imbued

33 "Hebrew Rabbi Again Before City Stipendiary."

34 "Hebrew Rabbi Again Before City Stipendiary." The rest of the account of the trial of Rabbi Levitt comes from this report.

with no little Baptist theological reasoning. "Personally, I believe in the fullest respect for every person of every sect, and if their opinions are in conflict to the general statutes of the land they are simply martyrs to their faith if they are convicted."

While one might have seen that the writing was on the wall for Rabbi Levitt, O'Hearn continued with the defense case by calling the shochet to testify through his interpreter. Rabbi Levitt endeavored to explain shechita, indicating that it had been practiced by Jews for three thousand years and that it was a quick and merciful way of killing. The magistrate and the prosecutor each intervened with a note of what was for them no doubt seen as jocularity. When Rabbi Levitt explained the ancient lineage of the Jewish method of killing, Fielding interrupted.

> The court laughed at this, having anticipated such a statement, and the stipendiary remarked: —"Mr. Interpreter, tell him he will have to cut out history and give us facts."
>
> Mr. Murray: —You might ask him how old he is?"

The white haired old man from Russia, testifying through an interpreter in a Canadian courtroom, asked by his attorney to explain how he had gone about killing the heifer on March 19, 1913 in accordance with Jewish law, was ridiculed and belittled by the good Protestant magistrate and the prosecutor, who had no interest in history, let alone Jewish religious belief and practice. In cross-examination, the levity and ridicule continued. When Rabbi Levitt explained to Murray that the religious laws of kashrut and shechita meant that all the blood was removed from the meat and that this was a health concern, he was again subjected to ridicule by the prosecutor and the magistrate.

> Mr. Murray (to the interpreter): —Then ask him how he accounts for all the Gentiles being alive?"

When the SPC prosecutor continued in the same vein and asked the rabbi to identify the name of the poison in the blood, "the stipendiary quietly remarked: – 'Blood poison.'"

O'Hearn closed his case and reiterated that the law only prohibited unnecessary suffering and cruelty. The defense barrister insisted that the Jewish method was not intentionally malicious as required by the language of the law that had to be interpreted according to all the circumstances, including religious belief and practice. Once more, the magistrate had an answer, invoking the criminalization of the practice of suttee in India or of religious customs in other countries that allowed a man to cut off a piece of flesh of a living animal "as it was walking about." The idea that these were invoked as direct and convincing analogies with the facts of the case and with the practice of shechita provides insights into the mind of the magistrate about exotic religious practices, in which category he clearly placed the Jewish method of slaughter. The direct comparison with the immolation of the Hindu widow in suttee with the killing of a heifer in a Halifax slaughterhouse not only provides a clear indication of the judicial mindset and understanding of Judaism abroad at the time, but it demonstrates the power of a combined legal positivism, religious ignorance and bias, and a welfarist attitude underlying the dynamics of social reform in Halifax in 1913, of which anti-cruelty and the SPC were but a part.

The specifics of Jewish laws of kashrut arose once the defense had rested. Fielding appears at this late stage to have engaged with O'Hearn's argument concerning the religious based practice of slaughter and dietary rules, and the relevance of those considerations in interpreting the anti-cruelty statute. The magistrate asked the translator, Webber, about the compulsory nature of kashrut. He inquired whether it was true that this was a religious obligation, since he could find no evidence of a divine command in either the Book of Leviticus or in Deuteronomy. He asked Webber whether it was not the case that many Jews ate pork despite clear religious prohibitions. Webber replied that anyone who ate pork was not living in accordance with Jewish law. He added that "thousands of people would starve if they could not have meat killed by the Jewish method."

Webber's answers were in keeping with Orthodox practice and belief in Halifax in 1913. But on the shechita question in particular, his response reflected the true nature of the case and its possible implications for the Jewish community. This was lived and experienced as a "major attack on Jewish freedoms." There were not thousands of Jews in Halifax or in Nova Scotia. Webber, a key representative of the local Jewish community, who had sat in the courtroom and closely observed the proceedings as Rabbi Levitt's interpreter and now as an erstwhile expert witness, testifying on the personal initiative of the magistrate and answering his inquiries into Jewish theology, obviously understood Murray's case as a frontal attack against Jewish religious practice. This was one case, prosecuted by the SPC on the facts of a single killing of a heifer, grounded in a narrow technical legal debate about conflicting meanings of "unnecessary" in the relevant statutory language as asserted by prosecution and defense. But this was more than that. As their case and evidence made clear, Murray and the SPC opposed all slaughtering that did not begin with stunning the animal. They opposed shechita as a matter of principle and this one case was meant to serve as the basis for pursuing a wider ban on the practice. Sam Webber had heard Fielding's comments about the law and religious martyrdom awaiting those who violated it. He realized that the consequence of the case against Rabbi Levitt, in the port of Halifax, Nova Scotia, could adversely affect the rights and religious practices of thousands of Canadian Jews.

Fielding adjourned the case to allow the defense to produce religious authority for the practice of shechita and to allow Murray to present what he claimed were two older Massachusetts cases of a similar nature to the present prosecution.

Criminalizing Jewish Practice in Nova Scotia

When the hearing in the Levitt case resumed, Murray informed the court that the Massachusetts cases on which he had sought further details were ordinary instances of animal cruelty and did not involve attempts to criminalize shechita. But the absence of precedent did not deter the magistrate or adversely affect the prosecution's case. On April 21, 1913, the local newspaper reported that "for the first time on the American continent a Jewish rabbi has been fined for killing an animal according to the Jewish belief, namely by cutting its throat and letting it bleed to death."[35]

In convicting the rabbi and imposing the fine Fielding appeared to try to minimize the blow by stating that his decision did not set a precedent. It was a decision on the unique facts of the particular case. But what he gave with one hand, he took back with the other. There was little evidence of the Christian mercy that was otherwise said to characterize Fielding's time on the bench in Halifax. He accepted the evidence offered by the SPC that stunning before killing was by far more humane than killing without stunning, i.e. the Jewish method. The consequence of this would be less cruelty to animals. But there were other consequences, alluded to by Webber, and acknowledged by Fielding. "If there are two methods by which (killing animals for food) can be accomplished and one is more humane than the other, I think that the more

35 "Jewish Rabbi Fined by the Stipendiary," *Halifax Herald*, April 22, 1913. Quotations from the judgment come from this account.

humane method should be followed. If the least [*sic*] humane method is followed it is unnecessary cruelty. It has been suggested that to convict would deprive a certain portion of the community of food. That may be."

Magistrate Fielding was content that he was fulfilling his judicial duty in administering the law. A conviction on the facts was inevitable whatever religious or conscientious scruples might have informed the rabbi's actions. He contradicted his opening remarks about not wishing to set a precedent when he declared that his decision about the unnecessary cruelty involved in the three-thousand-year-old Jewish method of killing was "really settling a principle." He recognized the wider and deeper implications of his decision by delaying the payment of the fine and costs for a week in order to allow O'Hearn to consider an appeal.

In the meantime, the conviction of Rabbi Levitt on the charge of animal cruelty had other immediate consequences. The SPC, through its secretary and lead counsel, Murray, announced a province-wide campaign in Nova Scotia against any butcher, Jewish or otherwise, who killed any animal large or small without stunning it first.[36] A similar plan to introduce compulsory stunning was envisaged for the neighboring province of New Brunswick. The SPC clearly saw the Levitt conviction as establishing a legal precedent the effect of which was not only that the Jewish method of slaughtering had been declared inhumane and cruel, but more broadly that stunning was now the legally recognized humane mode of killing animals for human consumption. The victory in convicting the Halifax rabbi had served the broader normative goals of the SPC in protecting animals from unnecessary suffering. The idea of humane slaughter and of stunning animals before killing them had been central to organizations dedicated to the prevention of cruelty to animals for many years, and that view had been vindicated in the Halifax courtroom.

The Jewish community in Halifax and in Nova Scotia at the time was not in a position to mount an effective campaign against the introduction of legislation banning slaughter without stunning. Jews were few in number, lacked financial resources, and could not be considered to possess political power in the province. But the widespread practice of killing without stunning throughout Nova Scotia may provide an explanation for the SPC's failure to mount a legislative campaign, and to rely on prosecution instead. There would have been little will in the provincial legislature for members to support a bill that would adversely affect many of their farming constituents who practiced

36 "A Campaign Against the Butchers," *Halifax Herald*, April 28, 1913.

traditional slaughter without stunning. By proceeding against a weak minority, and prosecuting the Halifax rabbi and shochet, the SPC may well have selected the least politically problematic path to achieving its broader goals without risking legislative defeat in largely rural Nova Scotia.

But Murray understood the case in more specifically Jewish terms. For the prosecutor and defender of animals, the Halifax conviction of the shochet with its blanket condemnation of shechita, was a case of jurisprudential moment. Nova Scotia, its SPC and its judicial organs, had set an important precedent and shown the way. Animal protection societies from all parts of Canada and the United States had written to Murray seeking details of the case. "The Jewish method of slaughter has been discussed a great deal in England and different parts of America but no cases have been brought under the statute. I feel that in some American cities it would be difficult to secure a conviction, but Halifax is particularly fortunate in having a fearless magistrate who will deal with such cases."[37]

Debates about shechita in terms of its cruelty had been abroad in both England and the United States. The Nova Scotia SPC was aware of the international circulation of ideas on this topic. Halifax may have been at the edge of Canada in the physical and geographical senses, but it was at the heart of wider debates in animal welfare circles. Murray put the case more strongly. Nova Scotia, through its fearless magistrate Fielding, was at the cutting edge of using legal means to criminalize and prohibit the Jewish method of slaughter. Less than a year before the Halifax prosecution, the RSPCA in England had again attacked shechita. The practice "has been condemned many times by practical people who have studied the subject, and one can only wonder, with regret, that so enlightened and cultured a race can still consider it necessary to retain among their religious tenets one which in these days of higher civilisation has been condemned as cruel and unnecessary."[38]

Judges in some American cities lacked the jurisprudential fortitude of the stipendiary magistrate in Halifax. One does not have to dig too deeply into Murray's statement to detect the presence of an antisemitic motif in the SPC's position. No doubt the significant factor at work in those American cities where the judges lacked Nova Scotian fearlessness was a pernicious and overwhelming Jewish influence that served to protect unnecessarily cruel and inhumane slaughter practices under the guise of religious belief. The true

37 Ibid.

38 "An Interesting Correspondence on Humane Methods of Slaughtering," *Animal World*, May 1912, 97.

nature, as a moral and legal principle, of shechita had been put into the light of day through the courage of the Nova Scotia SPC in bringing charges against Rabbi Levitt, and by Magistrate Fielding in convicting him.

While the SPC used the conviction of Rabbi Levitt as a springboard for a broader campaign against all slaughter of animals conducted without prior stunning, and as the SPC in Halifax and perhaps elsewhere perceived the case as establishing an important precedent criminalizing the Jewish method, O'Hearn and others were taking steps to ensure that an appeal could be brought to overturn the conviction and to offer some form of legal protection for the practice of shechita. The wider implications of the single conviction of the shochet were of key importance to Canadian Jewry.[39] The *Canadian Jewish Times* (*CJT*) took the lead. Founded by Lyon Cohen, a prominent figure in Montreal and Canadian Jewry,[40] and his co-publisher, S.W. (Sam) Jacobs, the *CJT* as the voice of the English-speaking establishment of Canadian Jewry, took a keen interest in the case. Jacobs, like Cohen, was a leader of the Montreal and Canadian Jewish communities. He served as head of the principal Jewish charity in Montreal, the Baron de Hirsch Institute, and practiced as a prominent lawyer in the city. In 1917, he became the first Jewish Member of Parliament from Quebec.[41] The *CJT* dedicated significant time and space to coverage of events in Halifax, especially following the conviction of the shochet.

In its early treatment, the paper had adopted an ambiguous attitude to the defendant himself, Rabbi Levitt. Based on the only evidence presented before the magistrate from Williamson and Gough, the *CJT* accepted at face value the statements that the heifer took fourteen minutes to die. For the editors, this was clear evidence that the killing for which Levitt was charged and convicted had not been carried out in accordance with the laws of shechita. In offering this preliminary analysis, they were attempting to limit the legal significance of Magistrate Fielding's decision by arguing that this was a case of animal cruelty that had little, if anything, to do with the correct implementation and practice of the rules relating to shechita. In doing so however, they were willing to throw Rabbi Levitt under the legal bus. "Another point which has to be considered

39 "Rabbi Criticizes Nova Scotia Judge's Lack of Knowledge," *Winnipeg Tribune* (MB), April 30, 1913.

40 "Lyon Cohen, Montreal," Hart, *The Jew in Canada*, 339; Bernard Figler, *Rabbi Dr. Herman Abramowitz, Lazarus Cohen, Lyon Cohen* (Gardenvale, PQ: Harpell's Press Cooperative, 1968).

41 "S.W. Jacobs, K.C., M.P., Montreal," in Hart, *The Jew in Canada*, 379; Bernard Figler, *Biography of Sam Jacobs* (Gardenvale, PQ: Harpell's Press Cooperative, 1959 and 1970).

(again if the evidence of the prosecution was correct), is whether this particular shochet thoroughly understands the duties incumbent upon him in the carrying out of the slaughter of animals by the behests of Jewish law."[42]

This was a line of argument that would be repeated in a letter from M.B. Steine to the *Halifax Herald* just before the appeal in the Levitt case was heard. Steine, a leading member of Jewish charities in Montreal, and a key actor in early Canadian Zionism, offered a detailed summary of Jewish dietary and slaughtering laws for the readers of the Halifax newspaper. He set out, as O'Hearn had attempted to do in Rabbi Levitt's defense, the ways in which the laws of shechita at their core respected and honored the animal and rejected any cruelty in the slaughtering process. But before his detailed analysis of Jewish dietary laws, Steine launched an unforgiving attack on Levitt's credentials. He sought to limit the case to its facts by arguing that Rabbi Levitt was incompetent and the killing in question had not been carried out in accordance with Jewish law. But he went further, by attacking Levitt's credentials as a religious leader in Halifax and reducing him to the status of "a veterinary slaughterer." "I wish to say that Abraham Levitt, the man who slaughtered the heifer, which resulted in his arrest for cruelty to animals, is no Rabbi, and has no more right to that title, accorded to him by The Herald, than Veterinary Surgeon Gough would have the right to the title of archbishop."[43]

Steine's pronouncement no doubt would have come as a surprise to the members of the Halifax Jewish congregation who had appointed him as their religious leader, or rabbi. Steine's unqualified assertions also would have come as a surprise to many other Jews, especially at this time. The title in question was not as a matter of law or practice necessarily one accorded as part of a formalized and hierarchized educational structure leading to professional qualifications and recognition. While Jewish theological institutions did exist at the time, a limited formal understanding of the honorific "rabbi" was not etched in stone as Steine contended. His attempts to create a clear distinction perhaps between a rabbi and a shochet, giving more status to the former, was intended to diminish Levitt's position. In doing so, he transformed the question into one of personal incompetence, and distanced the Jewish congregation and Jewish law from this isolated instance. This attack was ill-founded and ill-conceived, and perhaps could have served only to detract from the more central point

42 "The Halifax Shechita Trial," *Canadian Jewish Times* (hereafter *CJT*), April 25, 1913.

43 "The Provisions of the Hebrew Law for the Prevention of Cruelty," *Halifax Herald*, June 15, 1913.

relating to Jewish laws of kashrut and shechita, and the impending appeal in the case.

Fortunately, despite its early doubts about Rabbi Levitt's competence, the *CJT* quickly changed focus and gave its primary attention to the potential for harm arising from Fielding's decision. "If, however, the shochet did perform his duties satisfactorily, and the prosecution was merely meant to combat Jewish shechita, then the matter is no more a local one affecting the small Jewish community at Halifax, but is one in which the interest of Jews in all parts of Canada is concerned."[44] The newspaper issued a call for steps to be taken, should an appeal to be launched, to bring before the court convincing evidence of the merciful nature of shechita. Failure to do so, and to allow the decision of the magistrate in Halifax to be affirmed, "may be fraught with serious consequences to Judaism in Canada."[45]

The *CJT* continued to hammer home the point, highlighting not just the significance of the case for Canadian Jewry, but underscoring the meager resources, financial and theological, besetting the Halifax congregation. The Jews of Montreal, with their larger numbers, greater capacities, and their larger stake in the outcome of the attempt to criminalize shechita, had a duty to come to the aid of their fellows on the Atlantic coast. Rabbis and scholars in particular had an obligation to step forward to help Halifax Jewry establish before the courts a proper understanding of Jewish law and practice.[46] The *CJT* did not shy away from attributing antisemitic motives to the SPC's attack on Rabbi Levitt.[47]

On June 2, 1913, Louis Webber, as President of the Baron de Hirsch Benevolent Society of Halifax, wrote to the *CJT* formally seeking assistance from the Montreal Jewish community in "a matter which concerns all the Jewish people in the Dominion."[48] Among those who heard the appeal was Rabbi Simon Glazer. Throughout his time in Montreal, Rabbi Glazer was closely involved with all aspects of kashrut, including the many controversies and conflicts surrounding attempts to unify the regulation and licensing of kosher butchers in that city. Glazer travelled to Halifax at his own expense and served as the expert rabbinical witness for the appellant on questions relating to kashrut and shechita.[49]

---

44 "The Halifax Shechita Trial."

45 Ibid.

46 "Halifax Shechita Trial," *CJT*, May 2, 1913.

47 "The Shechita," May 16, 1913.

48 "An Appeal from Halifax," June 6, 1913.

49 Ira Robinson, "Rabbi Simon Glazer: A Rival for the Chief Rabbinate," in *Rabbis and Their Community: Studies in the Eastern European Orthodox Rabbinate in Montreal, 1896–1930*

County Court Judge W.B. Wallace heard the appeal, a trial de novo, from the stipendiary magistrate's verdict in June.[50] Wallace had an outstanding reputation as a jurist and as a social progressive. He was a noted advocate for improved housing conditions as a way of combating the social causes of crime, as well as being the first judge to sit on the newly established Halifax Juvenile Court.[51] He occupied a position of some status in Nova Scotia's legal establishment. A Roman Catholic, he was a keen supporter of the Sisters of the Good Shepherd and their work in reforming juvenile delinquents and prostitutes.[52] The same year as the shechita trial he was one of three judicial members of a royal commission of inquiry into land dealings of the provincial attorney general.[53] Murray again appeared for the SPC and O'Hearn represented Levitt.

The evidence of the SPC was a replication of that presented before the magistrate. Williamson, the SPC inspector, and Dr. Gough, the veterinary surgeon, repeated their testimony about what they had witnessed at the slaughterhouse in March when Rabbi Levitt had killed the heifer. Gough reiterated his expert scientific opinion that what he had seen was clear evidence of the prolonged suffering of the heifer and that the suffering was directly attributable to the Jewish method employed by Rabbi Levitt. The prosecution called additional witnesses, each of whom had experience in slaughterhouses in Halifax. Each testified to his preference for the "Gentile" method of slaughtering. J.H. Fielding of McLeod's slaughterhouse with sixteen years' experience; Sidney Tucker of Leaman's slaughterhouse, with seventeen years in the trade; Howard Gross, who was present when Rabbi Levitt had killed the heifer; James Shortell, owner of the slaughterhouse where the heifer had been killed; all testified that the stunning method caused less suffering to the animal.

(Calgary: University of Calgary Press, 2007), 35–55; King vs. Levitt, Evidence, ZD Halifax, Levitt, Alex Dworkin Canadian Jewish Archives, Montreal (CJ Archives).

50 The Notice of Appeal was presented to the County Court in Halifax on April 29, 1913, and both parties filed on May 13 and 14. District Court Number One (Halifax County), Cause Book from A13816–18174, Case 17886 Between the King And Abraham Leavitt [*sic*], Nova Scotia Archives. The actual County Court Case File, 17886, is missing from the relevant file box at the Nova Scotia Archives, RG38 291.

51 Boudreau, *City of Order*, 24, 60, 85. The origins of the Juvenile Court are described in Timothy T. Daley, "A History of the Nova Scotia Family Court," in Kimball, *The Bench*, 144–147.

52 Boudreau, *City of Order*, 254, n. 46; The Monastery of the Good Shepherd, operated by the Sisters of Charity "for delinquent, Roman Catholic girls and young women" had been operating since 1885. Lafferty, *The Guardianship of Best Interests*, 39.

53 Girard, "The Supreme Court of Nova Scotia, Responsible Government, and the Quest of Legitimacy," 456.

Two elements, completing the tale of the Halifax shechita trial from the SPC perspective, emerged from Williamson's testimony.[54] Williamson in his description of traditional slaughter practices he had observed and compared with what occurred on March 19 at Shortell's, for the first time referred to the Gentile method of killing. The opposition of the humane stunning method and the inhumane and cruel method without stunning of shechita that had informed the first trial and the ultimate decision by Fielding in April, was now presented in a more plainly religious juxtaposition, highlighting the antisemitic bases of the prosecution's case. While some of the comments by Fielding and Murray at the first trial had clearly carried a sub-text of mocking antisemitic stereotypes, Williamson's testimony gave full voice to the opposition that was clearly at play in the SPC prosecution, the humanity of the Gentile method and the cruel depravity of the Jewish method.

The second new element to emerge from Williamson's testimony, and one that buttresses the argument for the presence of a strong presumptive element of antisemitism in the SPC's pursuit of the case, is that the inspector's presence at the slaughterhouse on that day was neither routine nor accidental. "This was then the most important case I had. I was delegated by Mr. Murray to witness this. I was eager to get all the particular points in this case."

We know from the original trial that Dr. Gough had been asked to be present at the slaughterhouse in the company of Williamson as an independent expert witness.[55] The SPC had targeted Rabbi Levitt and the Jewish method of slaughter in a decisive and intentional fashion. Murray, the secretary and attorney of the Society, tasked Williamson with the case, and a designated expert witness supported him.[56] There can be little doubt given Williamson's contrasting of the Gentile and Jewish methods of slaughter, together with the clear targeting decisions made by the SPC, that the prosecution of Rabbi Levitt was more than a part of a broad campaign by the Society to end non-stunning in the province's abattoirs.

O'Hearn again sought to wage his argument at two different, but complementary and mutually reinforcing, levels. He offered evidence about shechita, its

54 All references are from the transcript/summary found in the CJ Archives, King vs. Levitt, Evidence, ZD Halifax.

55 Shortly after the trial, Gough was appointed an official agent for the SPC. Nova Scotia Society for the Prevention of Cruelty Fonds, MG 20 vol. 517, Minute Book, 1907–1925, August 13, 1913, Nova Scotia Archives.

56 Williamson resigned from his position as an SPC inspector effective September 1, 1913. Minute Book, 1907–1925, August 7, 1913.

rules and the practices associated with it, with a particular focus on the prohibition of cruelty, and the demand for mercy (*rachamim*) in Jewish law. He offered further expert scientific evidence to establish that the Jewish method was inherently humane and did not cause unnecessary suffering. He undermined Gough's testimony and his contention that the thrashing he saw was evidence of pain by bringing competing proof that the animal, once its throat had been slashed, had no central nervous activity that could result in pain.

The testimony on Jewish law and the method of slaughter came not from the white haired, non-English speaking Rabbi Levitt, or his translator, but from the far more authoritative, articulate, and experienced Rabbi Simon Glazer. As Robinson points out, a vital distinguishing characteristic of the Lithuanian-born Glazer was that he was a fluent English-speaker, who could and did write and publish books in English.[57] He brought to the Halifax courtroom gravitas, erudition, and experience as a rabbi, expert in kashrut and shechita, and the ability to articulate in an unmediated fashion the subtleties and complexities of Jewish law to the Nova Scotia court.

After establishing his theological credentials, Rabbi Glazer put paid to any doubts about Rabbi Levitt's qualifications as a shochet and a rabbi for the Jewish congregation in Halifax. He testified that in his fifteen years of rabbinical experience he had witnessed the slaughter by shechita of thousands of animals, and that when done according to Jewish methods, the killing of an animal could not result in pain or suffering. This was then supported by secular testimony led by the counsel for Rabbi Levitt.

O'Hearn examined Dr. Howard McFatridge, like Gough a graduate of the Ontario Veterinary College. McFatridge testified in no uncertain terms, and in direct opposition to his colleague Gough, that once the shochet cuts the animal's throat in the prescribed fashion "the animal is unconscious of pain. The struggling and kicking is a reflex action." More importantly, O'Hearn called Dr. David Fraser-Harris. The witness held a doctorate in medicine from the University of Glasgow and a further doctorate from the University of Manchester. He had completed postgraduate study in Berne, Jena, and Zurich. He was the Professor of Physiology at Dalhousie University and was an international expert on the physiology of nerves.[58] He testified without hesitation or prevarication that once "all the blood vessels in the neck are severed

57 Robinson, "Rabbi Simon Glazer," 36–37.

58 "David Fraser Fraser-Harris, D.SC, M.D.," *Proceedings of the Royal Society of Edinburgh* 57 (1938): 411–412.

consciousness vanishes from the brain in about five seconds. As to the convulsive period, this is not indicative of pain." He asserted that if the testimony from Gough and Williamson about what had occurred in Shortell's and how Rabbi Levitt had killed the heifer were true, the animal was rendered unconscious and suffered no pain.

The case presented by Murray remained largely unchanged from the evidence presented before Magistrate Fielding. With a few non-experts who testified about the preferability of stunning animals before killing them, the Gentile method, Murray's case rested on the description proffered by Williamson and the testimony of Gough about the suffering of the heifer. O'Hearn had countered with expert testimony from a qualified veterinary surgeon that openly contradicted Gough's position, and evidence from Rabbi Glazer about Jewish law and the practice of shechita. From an internationally recognized scientist and Professor of Physiology at Dalhousie University, he obtained direct proof that the Jewish method of slaughter could not, and did not, result in the infliction of cruelty, pain, or suffering as a matter of physiological science. While the themes remained the same from the April proceedings, this time the defense lawyer had mounted a clearer and stronger defense of shechita from both the theological and biological perspectives. Judge Wallace did not wait to render his decision. After final arguments from Murray and O'Hearn, the judge quashed the conviction and acquitted Rabbi Levitt. "Court decides in favour of deft. and orders his discharge."[59]

The judge was careful to avoid overt criticism of the magistrate and instead focused on the different evidence presented to him, placing strong emphasis on the scientific proof from Professor Fraser-Harris not available to the magistrate.[60] This evidence from an internationally renowned expert in the physiology of nerves established beyond any doubt that the Jewish method of killing was humane and perhaps, following Professor Fraser-Harris' argument, scientifically established as being more humane than the stunning method. The support of Rabbi Glazer, the Montreal Jewish community, and the *CJT*, combined with the fuller and more complete presentation of argument and proof before Judge Wallace by O'Hearn, led to a great victory for Rabbi Levitt, the Halifax and Canadian Jewish communities, and to a resounding defeat for the SPC and lawyer Murray. No longer could the conviction of the Halifax rabbi

59 1998-019/147, District Court Number One (Halifax County), Criminal Proceedings Book, 1900–1917, Nova Scotia Archives.

60 "Conviction in Jewish Cruelty Case Was Quashed," *Halifax Herald*, June 16, 1913.

serve as a springboard for a broader, pan-provincial or even international attack on shechita, or on non-stunning slaughter more generally. Instead the SPC had been hoist with its own petard when the judge found that the evidence from eminent religious authorities and leading scientists established the humanity of the Jewish method of slaughter. As Judge Wallace said, "I am convinced that the Jewish method is as equally humane as the Gentile, and strong scientific testimony has been produced to show that it is more so."[61]

It is regrettable that Wallace repeated the taxonomic juxtaposition of the Gentile and the Jewish method in his reasons, but the judgment did remove any legal buttress for the SPC's antisemitic attempt to criminalize the practice of shechita. But this account does not tell the full story of the fate of the Jewish method of slaughter before the criminal justice system, nor can it even begin to approach a recounting of the broader social, political, and legal context in which the Halifax shechita trial must be understood.

Anti-Shechita Agitation in Context

That the Halifax shechita case attracted national and international attention is hardly surprising. The most influential Jewish community in Canada, in Montreal, took a role in the person of Rabbi Glazer who led the fight back against the conviction of Abraham Levitt. The *CJT* publicized the case to a national audience and, despite some minor kerfuffle within the Montreal Jewish community, fought back publicly against the possible national implications of a decision confirming the conviction.[62] A sense of triumphant relief was evident in the paper's reaction to the acquittal.

> In the Halifax Shechita Trial, we can proudly say that it was the "Canadian Jewish Times'" fight. By giving such publicity to the trial and showing the possible dangers of the situation, we aroused the communities of the Dominion, to feel that a great injustice had been done to the Halifax community, which could result in endangering the position of Jews in the Dominion. ... We cannot leave this matter without referring to the assistance given to our Halifax brethren by Rabbi Simon Glazer, who journeyed to Halifax and gave evidence at the hearing of the appeal.[63]

61 Ibid.

62 "An Appeal from Halifax,"; "The Duties of a Communal Paper," June 13, 1913.

63 "Editorial," June 20, 1913.

The case on its own was of central importance to Jewish communities throughout Canada and the *CJT* reaction reflects this. As was noted by Samuel Webber in April 1913, the availability of kosher food was a core element to the religious identity of the Orthodox Jews of the Halifax community and the same obtained throughout the country and elsewhere.[64] The prosecution of Rabbi Levitt and his conviction endangered the lives and the identities of Jews in Halifax and beyond in its potential effects. Moreover, it was clear that anti-Jewish motives informed, in part at least, the legal proceedings led by the Nova Scotia SPC.

Beyond the problematic nature of any prosecution of a shochet under animal cruelty legislation and its direct and indirect antisemitic resonance, Canadian Jews at the time confronted other vital political and judicial instances of importance to their existence and identity as emancipated equal citizens. In Montreal, at the same time as the decision acquitting Rabbi Levitt was being handed down by Wallace, the *CJT* highlighted the ongoing controversies and debates surrounding the vexed Jewish School Question in the city, and in particular delicate negotiations over the hiring of Jewish schoolteachers by the Protestant School Board.[65] In Quebec City, the infamous case of *Ortenberg v Plamondon* pitted the tiny Jewish community against virulent Roman Catholic ultramontane attacks, verbal and physical, including the blood libel.[66] The defense of Jewish rights in Quebec City was led by Sam Jacobs, supported by the much larger Jewish community of Montreal, a community that had come to Rabbi Levitt's aid.[67] Montreal Jewry, in 1913, was fighting battles on several fronts, of which the Halifax shechita case was one, in its efforts to concretize the promise of equality and citizenship that had brought so many European Jews to North America. Across the Atlantic, in Ukraine, as the Quebec City trial was taking place, Mendel Beilis stood accused by Russian authorities of

64 M. Hyamson, "Another Word on the Dietary Laws," *Jewish Quarterly Review* 9, no. 2 (1897).

65 David Fraser, *"Honorary Protestants:" The Jewish School Question in Montreal, 1867–1997* (Toronto: University of Toronto Press/Osgoode Society for Canadian Legal History, 2015).

66 1914 CarswellQue 71; 24 B.R. 385; 35 CLT 262 (1915).

67 David Rome, *The Plamondon Case and S.W. Jacobs*, 2 vols. (Montreal: Canadian Jewish Congress National Archives, 1982); Sylvio Normand, *L'affaire Plamondon: un cas d'antisémitisme à Québec au début du XXe siècle*, 48 Cahiers de Droit 477 (2007); Constance Backhouse, "Anti-Semitism and the Law in Québec City: The *Plamondon* Case, 1910–15," in *Transformations in American Legal History: Law, Ideology, and Methods, Essays in Honor of Morton J.* Horwitz, *Volume Two*, eds. Daniel W. Hamilton and Alfred Brophy (Cambridge, MA: Harvard Law School/Harvard University Press, 2010).

the murder of a Christian child, a prosecution grounded solely in the anti-Jewish ritual murder myth. European and world Jewry stood as one against this pernicious blood libel, and the direct parallels between the legal confrontation over the blood libel and Jewish demands for recognition as equal citizens in Quebec City and Kiev did not pass unnoticed by Canadian Jews.[68]

The charges against Abraham Levitt would and did resonate with these and other instances of antisemitic juridified violence occurring in several jurisdictions, each involving Christian mythologies of Judaism as a particularly cruel, murderous, and bloodthirsty religion.[69] The juxtaposition of the humane Gentile method of slaughter and the cruel infliction of unnecessary pain and suffering through the Jewish method reflected more than the direct comparison between the two signifiers Jewish and Gentile in the context of the prosecution of Abraham Levitt. The comparison and contrast between Jewish cruelty and Gentile humanity embodied much broader local, national, and international realities, imaginaries, and signifiers of antisemitism that were at play socially, politically, and juridically at the time. Halifax sat at the edge of Canada, on the shores of the wild Atlantic, and perhaps at the periphery of the British Empire, but discursively it was located at the center of an internationally situated acute contemporary jurisprudence and politics of anti-Judaism.

Jews situated at the center of the empire in London, and throughout the United Kingdom, followed the Halifax shechita trial in the pages of the *Jewish Chronicle* (*JC*). First published in 1841, the *JC* is the oldest continuously published Jewish newspaper in the world. Its influence on, and reach into, the British Jewish community cannot be underestimated.[70] Attacks against shechita in the name of the prevention of cruelty to animals continued in England in 1913 and after, making the Halifax trial of some current local, as well as international, significance.[71]

The *JC* carried brief coverage of the conviction and of Murray's pronounced intentions to mount a campaign against all slaughter without

---

68 David Fraser, "The Blood Libel in North America: Jews, Law, and Citizenship in the Early 20th Century," *Law and Literature* 28, no. 1 (2016).

69 David M. Friedenreich, *Foreigners and Their Food: Constructing Otherness in Jewish, Christian, and Islamic Law* (Berkeley: University of California Press, 2011); David Biale, *Blood and Belief: The Circulation of a Symbol Between Jews and Christians* (Berkeley: University of California Press, 2008).

70 David Cesarani, *The Jewish Chronicle and Anglo-Jewry, 1841–1991* (Cambridge: Cambridge University Press, 1994).

71 "Mr. Galsworthy and Shechita," *Jewish Chronicle*, April 18, 1913; "Agitation Against Shechita in Nova Scotia: A Shochet Charged with Cruelty," ibid.

stunning. For the London paper's correspondent, "The campaign is not specifically anti-Semitic, but, if persisted in, it will have serious effects for Orthodox Jews who will be unable to obtain kosher meat."[72] While unwilling to attribute anti-Jewish motivations to the SPC, the London newspaper was conscious of the anti-Jewish impact should such a campaign succeed.

The *JC* later saw the upside of the Halifax shechita case following the decision of Judge Wallace to overturn Rabbi Levitt's conviction in terms that confirmed the humanity of the Jewish method of slaughter. In applauding the victory of the Halifax shochet, the paper may have overstated the technical legal place of the Wallace decision in writing that "against the windy declamation of anti-Shechita enthusiasts may now be set against the considered judgment of an important Canadian Court."[73] The *JC* underscored a key point on the thinking of English and Canadian Jews at the time. "The fact shows that Jews may expect fair treatment in this matter if only they bestir themselves and bring the truth to light."[74]

There is little doubt that the Jewish communities in Canada and England experienced the final decision in the Halifax case as a victory for Jewish rights and as proof that the system of English law, both at home and in the empire, was open to the equal treatment of Jewish religious belief and practice. Such a position was originally belied by the original course of the case before the Halifax magistrate, but a more carefully elaborated defense case, including the powerful combination of the scientific evidence of Fraser-Harris and the practical and theological influence of Rabbi Glazer's intervention, did demonstrate that the judicial arena could be, and was, amenable to the elaboration and defense of Jewish rights. As the *JC* rightly underlined, and as the experience of the Halifax shechita case, and other Canadian questions going on at the same time in Montreal and Quebec City amply demonstrated, an engaged and bestirred Jewish community was a key factor in legal success. But in England and elsewhere, the struggle between the Jewish community and advocates of the banning of shechita continued despite the verdict in the Halifax case.[75] Anti-cruelty advocates agitated for legislative reform to compel the stunning of all animals before slaughtering.[76]

---

72 "The Conviction of a Jewish Shochet at Halifax (Nova Scotia)," *JC*, June 13, 1913.

73 "A Victory for Shechita," *JC*, June 27, 1913.

74 Ibid.

75 "Attack on Shechita," *JC*, July 4, 1913.

76 "The 'Humane' Slaughtering of Animals," *JC*, July 11, 1913; "The R.S.P.C.A. and Shechita," August 1, 1913.

The trial in Halifax sat within a matrix of relationships, of knowledge and power, and of the circulation of ideas within the world of anti-cruelty organizations. The Nova Scotia Society for the Prevention of Cruelty was listed among the "Colonial and Foreign Societies" engaged in a corresponding relationship with the Royal Society for the Prevention of Cruelty to Animals (RSPCA) in London, along with similar bodies throughout the empire.[77] Murray had almost joyously reported upon Levitt's conviction in April 1913 that he had been immediately contacted by Societies across North America.[78] In the same statement, Murray specifically referred to a report from a committee of the House of Commons in England condemning slaughter without stunning. He was referring to the infamous Admiralty Report of 1904, under the Chairmanship of Arthur Lee, MP, a long-time supporter of the RSPCA and an ardent anti-shechita advocate.[79] That Committee had received extensive evidence on slaughtering practices, including the regionalized practice of killing without stunning, and the more widespread slaughtering of some types of animals as matter of course without stunning.[80] In particular it emerged that there was an almost universal practice of killing sheep without stunning throughout the United Kingdom. The Committee recommended "All animals, without exception, should be stunned, or otherwise rendered unconscious, before blood is drawn."[81] They urged the establishment of a system of public abattoirs to replace private slaughterhouses, and that a system of official supervision and inspection of all slaughtering practices and facilities be instituted in the public abattoirs. All slaughtermen were to be licensed by local authorities. These proposals constituted an important tightening of rules already in part at least in place in the legislative scheme governing slaughterhouses in England.[82]

Despite having received extensive testimony from representatives of the Jewish community, the Committee maintained its global recommendation on stunning, an idea which, if implemented, would have outlawed shechita. It chose to ignore or downplay the testimony of Sir Simon Montagu, President

77 Royal Society for the Prevention of Cruelty to Animals, *Ninetieth Annual Report, 1913* (London: RSPCA, 1913), 41, 87, 123.

78 "A Campaign Against the Butchers."

79 Report of the Committee Appointed by the Admiralty to Consider the Humane Slaughtering of Animals (London: HMSO, 1904); Arthur Lee, "The Meat Trade and Its Defence: A Criticism," *Daily Mail*, January 13, 1913; "Humanity in the Slaughter-House," July 1, 1913.

80 Minutes of Evidence Taken Before the Committee Appointed by the Admiralty to Consider the Humane Slaughtering of Animals (London: HMSO, 1904).

81 Report of the Committee Appointed by the Admiralty, 6.

82 Ibid.

of the London Board of Shechita. Instead it gave greater credence to the evidence presented by German officials from Lander and cities where shechita had been abolished.[83] Perhaps not surprisingly, an expert report commissioned by the Committee found that the Jewish method of slaughter was crueler than killing after stunning.[84] The Board of Deputies of British Jews had immediately formed a new Shechita Committee to combat this attack on religious practice.[85]

In addition to his invocation of the Admiralty Report, Murray made specific references to the districts of Baden, Saxony, and Bavaria in Germany and to Switzerland where the Jewish method was banned.[86] It was, according to Murray's own statement to the press, in Germany that shechita was the prime target for anti-cruelty activists for "the chief thing the Prussian societies are fighting is the Jewish method of slaughter."[87] Knowledge about issues of animal cruelty, but more specifically about shechita, circulated among Societies for the Prevention of Cruelty. Murray was familiar with prohibitions in parts of Germany and in Switzerland on slaughter without stunning, including the Jewish method. He was aware of these prohibitions against this form of slaughter before he commissioned Inspector Williamson and Dr. Gough to visit Shortell's slaughterhouse in Halifax to witness Rabbi Levitt killing the heifer. Murray and the Nova Scotia SPC participated in the international network of knowledge about bans on shechita. That knowledge must have affected the decision in Halifax, on Canada's east coast, to prosecute the rabbi of the small Orthodox Jewish community for animal cruelty.

It is doubtful that Murray and his Nova Scotia SPC colleagues were intimately familiar with the more deeply embedded cultural contexts of debates in Germany that had led to attempts, often successful, to ban shechita.[88] But they did have a cultural and theological familiarity with related ideas about the connections between cruelty and cultural degeneration. Those questions were at the heart of the social reform movement in which

---

83 Minutes of Evidence, 51 et seq.

84 M. Foster and Ernest H. Starling, Report on the Jewish Method of Slaughtering, in Report of the Committee Appointed by the Admiralty, 24–26, Appendix J.

85 London Committee of Deputies of British Jews, 54th *Annual Report,* 1904,14, 27.

86 "A Campaign Against the Butchers."; Robin Judd, *Contested Rituals: Circumcision, Kosher Butchering, and Jewish Political Life in Germany, 1843–1933* (Ithaca, NY: Cornell University Press, 2007).

87 Ibid.

88 Dorothee Brantz, "Stunning Bodies: Animal Slaughter, Judaism, and the Meaning of Humanity in Imperial Germany," *Central European History* 35, no. 2 (2002).

Murray, the SPC, and other members of Nova Scotia's Protestant elite were important actors. It does not require any great expansion of the imagination to believe that similar ideas about cruel Jews, fascinated with blood in their core rituals of circumcision and shechita, and perhaps in their more hidden practices of child murder, circulated both in Christian Germany and in Christian Canada and Nova Scotia. Ideology, theology, folk myth, and belief in civilized progress informed the international animal anti-cruelty movement of which Murray and the Nova Scotia SPC were a part.

In October of 1913, despite the acquittal of Rabbi Levitt, Murray reported to his colleagues at the Nova Scotia Society that he had received an invitation to attend the annual meeting of the American Humane Association in Rochester, New York, to deliver a paper on "Jewish Slaughtering." His fellow members of the executive denied his request for funding to attend the American gathering, citing "the shortness of finances."[89] For the international anti-cruelty movement, the issue of shechita had not been removed from the agenda notwithstanding the decision in Halifax before Judge Wallace. For anti-cruelty activists in the United States, shechita was still a significant issue, and current legal expertise in attempting to ban the practice was located in Nova Scotia. Despite the lack of financial support from the Nova Scotia body, Murray traveled to upstate New York in the autumn of 1913. He featured among British and Canadian anti-cruelty representatives who addressed the national convention of the American Humane Association.[90] The afternoon session of the convention featured "a friendly discussion" between Murray, who maintained his characterization of the practice as cruel and Louis Ginsberg [*sic*], Professor of Talmud and Rabbinics at the Jewish Theological Seminary and vocal defender of shechita, on the Jewish method of slaughter.[91]

The Nova Scotia shechita case was located in local, national, and international political, social, and legal debates about the place of Jews, and Jewish religious practices, both broadly conceived or more narrowly constructed to deal with the Jewish method of slaughtering animals. As the early coverage of the case in the *Halifax Herald* indicated, for all those involved this was a case of first impression, the first time a North American rabbi had been charged with

---

89 Minute Book, Meeting of October 1, 1913.

90 "Public Meeting in Interest of Humane Work," *Rochester Democrat and Chronicle*, October 13, 1913.

91 "Kosher Methods Attacked," *Rochester Democrat and Chronicle*, October 14, 1913; "Dr. L. Ginzberg, 79, Jewish Educator," *New York Times*, November 12, 1953.

CHAPTER 2

The Aberdeen Shechita Case of 1893: Criminalizing the Jewish Method in Scotland

The Aberdeen Prosecution

On August 30, 1893, James Littman, the rabbi and shochet of the small Jewish congregation of Aberdeen, Scotland, in the presence of the president of the community, Alexander Zamek, at the McIntosh slaughterhouse, killed a bullock. The local prosecuting authority, the Procurator Fiscal Robert Lamb, issued a summons against both Zamek and Littman charging them with animal cruelty offenses. It was alleged that the shochet and president had committed three separate acts of cruelty during the shechita, 1– they had tied the feet of the animal and 2– thrown it violently to the ground and 3– Littman had cut the throat of the bullock in an unskillful manner, causing the animal to struggle violently, break free for the restraints, and suffer an agonizing five minute period before death.[1] These alleged acts of cruelty violated the provisions of § 1 of the Cruelty to Animals Act (Scotland) and were liable to a punishment of up to three months imprisonment and a fine of five pounds and costs.[2]

The Aberdeen case of 1893 involves parallels with the Halifax prosecution twenty years later. As a matter of legal and factual substance, the cases involved charges some of which appeared to go to the practice of shechita itself, particularly accusations that killing without prior stunning was per se cruel, and at the same time elements particular to the facts of each instance, raising the question of the competence of the shochet in carrying out the act of killing.

---

1 "The Hebrew Congregation in Aberdeen: Mosaic Mode of Slaughter Objected To," *Aberdeen Journal*, September 28, 1893; Kenneth Collins, "A Community on Trial: The Aberdeen Shechita Case, 1893," *Journal of Scottish Historical Studies* 30, no. 2 (2010).

2 13 & 14 Vict c. 92 (1850).

The Aberdeen action characterized the acts preparatory to the actual killing of the animal by the shochet's single cut, restraining the bullock and placing it in the position where its throat was exposed to the blade, known as casting, as themselves cruel in addition to the cruelty of the actual slaying of the animal.

Socially, geographically, politically, and historically, the Jewish communities in New Scotland and in Aberdeen were strikingly similar. Aberdeen, like Halifax, was situated on the coast, isolated from the major centers of Scottish Jewry in Edinburgh and Glasgow, and far distant from the capital of United Kingdom Jewry in London. It was a small congregation of recent vintage. In both shechita cases, separated by twenty years, an isolated, largely impoverished immigrant Jewish community was targeted by authorities for a legal attack that threatened fundamental tenets of Orthodox faith and practice. One early historian of Scottish Jewry has remarked that for the Aberdeen Jewish community "so far as the nineteenth century is concerned, it has little or no history."[3] For a few days in the autumn of 1893, Aberdeen's Jews would be at the center of attention for Jewry across the United Kingdom.

As with Nova Scotia, the story of the Jews of Scotland is to a certain extent obscure and hidden in myth and half-told tales. Nathan Abrams, who has undertaken extensive studies of Scotland's dispersed Jewish communities says, "very little is known about Jewry in Scotland because of the small numbers, and what little evidence there is has often been handed down by word of mouth."[4]

The histories of the largest Scottish Jewish communities of Edinburgh and Glasgow have recently been the objects of more careful study.[5] Especially in the case of Glasgow, Jewish history has been placed in the social and political contexts of other immigrant groups to the large port city, and within the local situation of the identitarian divide between Protestant and Roman Catholic populations of Scotland's larger cities.[6] Victorian Glasgow experienced more generalized debates and controversies over public health and sanitation in relation to slaughterhouses and the quality of meat available in the city, like

3 A. Levy, "The Origins of Scottish Jewry," *Jewish Historical Society of England–Transactions*, 19 (1955): 157.

4 *Caledonian Jews: A Study of Seven Small Communities in Scotland* (Jefferson, NC: McFarland, 2009), 5; Kenneth E. Collins, *Scotland's Jews: A Guide to the History and Community of the Jews of Scotland* (Glasgow: Scottish Council of Jewish Communities, 2008).

5 Mark Gilfillan, "Jews in Edinburgh: 1880–1950," Ph.D. dissertation, University of Ulster, 2012; Ben Barber, *Jews in Glasgow, 1879–1939* (London: Vallentine Mitchell, 2007); A. Levy, *The Origins of Glasgow Jewry, 1812–1895* (Glasgow: printed by author, 1949).

6 William Kenefick, "The Jews and Irish in Modern Scotland: Anti-Semitism, Sectarianism and Social Mobility," *Immigrants and Minorities* 31, no. 2 (2013).

other large metropolitan areas at the time. No record seems to indicate that the Jewish method of slaughter caused any concern in this context of the emerging drive for greater control over all aspects of the killing of animals for human consumption.[7]

For smaller, regional cities, the work of Collins and Abrams has begun the process of constructing a more complete picture of the place of Jewish communities in Scottish history. One issue that has been prominent in the historiography of the Jews of Scotland is the marginalization of Scottish specificity under the category of British Jews, or the even more problematic Anglo-Jewish descriptor that is invoked as a common imperialized term meant to encompass England, Wales, Ireland, and Scotland. In analyzing the Aberdeen shechita case of 1893, it is important to maintain a careful distinction between those aspects of the story that are local, either Aberdonian or Scottish, and those with broader implications within Britain, or even simply in London. Just as the Halifax case some twenty years later would involve the local and the national (and the international), this iteration of attempts to criminalize shechita carried geographically specific elements and implications.

Jews began to arrive as a community in Aberdeen, as elsewhere in the United Kingdom, with immigrant waves from central and eastern Europe beginning in the 1840s. There had been an individual Jewish presence in the city for many years prior to this, especially in the form of Jewish medical students who attended the University of Aberdeen.[8] The community gradually established itself, and shortly before the prosecution for animal cruelty was undertaken, the eight families of the Jewish congregation of Aberdeen formally consecrated their synagogue, in two rooms of a building on the first floor of a house at 34 Marischal Street, with Rabbi Littman as religious head.[9]

James Littman was born in Meseritchi (Mezeritch) in what is now Ukraine, but was then described as the part of Poland ruled by Russia.[10] Littman had extensive training and experience as a *mohel* and as a shochet in both Poland and London. Collins speculates that it was his expertise in the Jewish method of

7 P.J. Atkins, "The Glasgow case: meat, disease and regulation, 1889–1924," *Agricultural History Review* 52, no. 2 (2004).

8 Kenneth G. Collins, *Go and Learn: The International Story of Jews and Medicine in Scotland* (Aberdeen: Aberdeen University Press, 1988).

9 "Consecration of a Jewish Synagogue," *Aberdeen Journal*, September 9, 1893; Abrams, *Caledonian Jews*, 21.

10 Certificate of Naturalisation, Duplicate Certificates of Naturalisation, Home Office: Immigration and Nationality Department, Series HO 334, December 3, 1902, National Archives. Littman is described as "a Hebrew minister."

slaughter that was the key to Littman's Aberdeen appointment. Aberdeen was a major center of beef production in the United Kingdom, and like other cities where slaughterhouses were situated, there was a record of concern over issues of adequate inspection of the processes in place, public health, and the hygienic operation of the establishments well before the 1893 case.[11] It was unlikely that the small, for the most part impoverished, fifty-person community, of whom only eight were full members of the congregation, and of whom only five had reasonably sound finances, would have had any real ongoing need or ability to pay, for kosher meat. On the other hand, the small Jewish community could supplement its income through the production of kosher beef for sale elsewhere. "We have to conclude that Littman came to Aberdeen precisely because of his shechita expertise in cattle as kosher meat from Aberdeen, then a center of a great meat production area, was greatly in demand and was to be sent to the London market."[12] Littman later left Aberdeen and moved to London where he worked as a shochet for the London Board of Shechita for many years.[13]

There had been concern about sanitation in abattoirs in Scotland for some time, as well as an important focus on slaughtering techniques.[14] This coincided in Scotland, as it did elsewhere, as meat production became industrialized and centralized, with a growing interest in humane slaughtering and stunning methods.[15] Where stunning and humane slaughter are on the social and political reform agendas, the question of the Jewish method cannot be far behind. As early as 1866, Edinburgh's *Scotsman* in reporting on the practice of shechita and the fate of animals killed by shochetim stated that "it is painful to think of the tortures they will endure. Even professional butchers, in the daily habit of taking away animals [*sic*] life, cannot stand to witness the Jewish method of slaughter."[16]

In Dundee, not too long before the prosecution in Aberdeen, a local committee of the Police Commission, accompanied by a veterinary surgeon and a representative of the local SPCA, had, following a complaint, attended

---

11 W.J. Simpson, Report to the Public Health Committee on the Inspection of Meat and the Need of Public Abattoirs in Aberdeen (Aberdeen, 1885).

12 Collins, "The Aberdeen Shechita Case," 77.

13 Census Returns of England and Wales, 1911, RG14, Piece 1146, National Archives, describing Littman as a "slaughterer shochet."

14 "The Proposed New Slaughter-House and Cattle Market for Dundee," *Dundee Courier and Argus*, January 13, 1872; "Opening of the New Slaughter-House, Leith," *Caledonian Mercury* (Edinburgh), April 10, 1863.

15 "Humane Slaughtering," *Daily Mail*, February 13, 1913.

16 "Butcher-Meat and the Jews in Edinburgh," March 27, 1866.

the local slaughterhouse to witness the killing of a bullock by the Jewish method. Members of the Jewish community of Dundee were in attendance.[17] The observers from the local government and anti-cruelty movement concurred in their assessment. "The general opinion of those present was that the ordinary rule was much to be preferred to the system adopted by the Jews."[18] Despite their obvious distaste for shechita and clear views that stunning was a far superior method for killing animals, Dundee officials took no action against the shochet. While stunning was preferred as the method, shechita in Dundee at least did not rise to the level of animal cruelty.[19] That was left to local law enforcement officials and the courts of Aberdeen.

The Aberdeen Prosecution: Scots Law and Jewish (Almost) Guilt

The trial of Alexander Zamek and James Littman before the local magistrate, or baillie, in Aberdeen Police Court was scheduled to take place on October 3, but the solicitor for the accused, William Lamb, moved for an adjournment on the grounds that the date was the Jewish religious holiday of Simchat Torah. Rabbi Littman did not speak English and a translator would have to be found before a fair trial could take place.[20] Before the commencement of proceedings, Alexander Zamek had issued a plea to Jews across the United Kingdom for funds to support the defense of the two Aberdeen men accused and more broadly, the rights of Jews to practice shechita. The *Jewish Chronicle (JC)* issued a declaration in support, noting the circumstances facing the newly established congregation. "The first occasion on which its Shochet entered upon his duties has been made a subject of persecution by the local Society for the Prevention of Cruelty to Animals."[21]

The organ of established British Jewry saw the case as one not of prosecution, but of persecution, led by the SPCA. The actual origins of the case against Rabbi Littman and Alexander Zamek are somewhat mysterious. The *JC* clearly indicated that the SPCA and its inspector were behind the

---

17 "Jewish Customs in Dundee: Their System of Cattle Slaughter," *Dundee Courier and Argus*, January 17, 1893; "The Jewish Method of Slaughtering Cattle," *Scotsman*, January 17, 1893; Nathan Abrams, "Jute, Journalism, Jam and Jews: The Anomalous Survival of the Dundee Hebrew Congregation," *Northern Scotland* 3 (2012): 96–97.

18 "Jewish Customs in Dundee."

19 Collins, "The Aberdeen Shechita Case," 77; "Killing 'Kosher' Cattle," *Aberdeen Journal*, October 5, 1893.

20 "The Jewish Mode of Slaughter," *Aberdeen Journal*, October 2, 1983.

21 *JC*, October 13, 1893.

summons. Their correspondent wrote, "The prosecution emanated from Mr. Strangeways, Inspector of the local Society for the Prevention of Cruelty to Animals, which I believe is largely composed of sentimental ladies."[22] At the trial itself, the Procurator Fiscal, the Scottish public prosecutor, who presented the case against the two Jewish officials, took sole responsibility for the charges.[23] Unlike the Halifax case, this was a legal action in its formal manifestations at least, involving the public power of the state making criminal charges against the rabbi and shochet and the president of the newly established Jewish congregation of Aberdeen. Whatever involvement the SPCA and its officials may have had, formally this was a case of the state bringing a penal proceeding against a small, relatively isolated Jewish community, and mounting what appeared to be a legal attack on the religious beliefs and practices of those Jews.

Zamek's pleas for support to the Jewish communities of the United Kingdom had been successful. As the *JC* indicated, for many Jews in Britain, this was more than just a singular and limited prosecution of two Jews in far northeast Scotland. This was understood as what Collins qualifies as "a British test case for shechita."[24] Elkan Adler, brother of the Chief Rabbi, was sent to Aberdeen to instruct the defense's solicitor and to act as an interpreter for Rabbi Littman.[25] Just as the Montreal Jewish community had seen the pursuit of Abraham Levitt as an attack on all Jews, and Rabbi Glazer had gone to Halifax to assist in the appeal, London and English Jewry understood the charges of animal cruelty leveled against the shochet and the leader of the congregation as embodying a broader assault on core beliefs and practices of the observant Jews of the United Kingdom.

The prosecution of the president and the rabbi/shochet of the tiny Aberdeen Jewish congregation finally took place on October 14, 1893 before a packed courtroom that could not hold all those who wished to attend.[26] The magistrate was Baillie George Kemp. Like most magistrates of the time, he was not legally trained. Instead he had been selected for this post as a prominent member of the local community. He was active in Aberdeen life and in

22 "The Shechita Question," October 6, 1893.

23 Collins, "The Aberdeen Shechita Case," 85.

24 Ibid., 86.

25 Hyman A. Simons, "Chief Rabbi Dr. Hermann Adler– A Period Study in Anglo-Jewry," *Judaism* 18, no. 2 (1969).

26 A verbatim account of the trial can be found in "The Hebrew Slaughter Case: Trial in Aberdeen-How 'Kosher' Meat is Killed," *Aberdeen Journal*, October 14, 1893 and in "The Shechita Question at Aberdeen: Trial and Acquittal of the Defendants," *JC*, October 20, 1893.

his Church. Kemp was a steamship company manager, and later an accountant and auditor. He was a longtime member of the City Council of Aberdeen, having served as Treasurer. He sat as baillie for only a short period. Like his Nova Scotian counterparts in the later prosecution of Rabbi Levitt, Kemp was a man of his time and a believer in social progress. He headed the gas and electricity departments in his time on the council, bringing technological advances to the Scottish city, and served as a supporter of temperance. He was a pillar of the Congregational Church and active throughout his life in the Sunday School movement.[27] The judge in the case, then, was a Victorian Christian advocate of temperance, of good works and social betterment.

In the formal court proceedings, James Littman was described as the Rev. James Littman, clergyman and Alexander Zamek as a moneylender. Zamek was a naturalized British citizen, described in his citizenship document as a watchmaker and jeweler.[28] His business "Messrs Zamek & Edelshain" was styled as wholesale dealers.[29] Following the trial, the local press consistently referred to Zamek as a "moneylender."[30] The 1901 Scottish census return confirms this characterization.[31] His death certificate from 1907 reported that he was a financial agent.[32] In Aberdeen in the last decade of the nineteenth century the rabbi and the moneylender stood accused of animal cruelty for an act directly related to Mosaic law.

> The charge against the accused was that in slaughtering a bullock according to the Mosaic law they cruelly ill-treated and tortured the animal. It was not alleged that this was done because the bullock was killed in the manner prescribed by the Jewish law, but because the operation was, the prosecution averred, conducted unskillfully.[33]

---

27 "Death of Treasurer Kemp: A Notable Citizen of Aberdeen," *Aberdeen Journal*, October 26, 1914.

28 Certificate of Naturalisation, Home Office, Duplicate Certificates of Naturalisation, Immigration and Naturalisation Department, Series HO 334, September 17, 1887, National Archives.

29 "Alleged Fraud in Aberdeen," *Aberdeen Journal*, August 14, 1891.

30 "A Money-Lending Case," *Aberdeen Journal*, July 18, 1896; July 22, 1896; "Extraordinary Money Lending Case," November 6, 1896.

31 Scotland Census 1901, Reels 1–446, General Register Office for Scotland, Edinburgh.

32 Calendar of Confirmations and Inventories, Commissary Clerk of Edinburgh, under the Sheriff Courts Act, 1876, December 6, 1907.

33 "The Hebrew Slaughter Case," *Aberdeen Journal*. The discussion is taken from this account.

Themes that are familiar throughout the legal history of attacks against shechita emerge. William Sim, a well-known Aberdeen lawyer, and Zamek's usual legal representative, acted for Littman and Zamek. Robert Lamb, advocate and Procurator Fiscal for the burgh of Aberdeen, led the prosecution case. Lamb, an active member of the Liberal Party, carried on a lucrative private practice in the city.[34] He was well known and particularly well regarded by local judicial officials.

> Many a Baillie was indebted to Mr. Lamb for valuable advice and guidance in the discharge of the duties of the bench, which in recent years have vastly increased on account of the growth of the city and the multiplication of petty offences which have been brought before the court under new Acts that come into operation.[35]

The prosecution in Aberdeen was quick to point out that this was not an attack on the Jewish mode of slaughter, but was a case of alleged animal cruelty based on the facts of the killing of the bullock on August 30, 1893 in the premises at Hutcheon Street, Aberdeen. That August 30 was the first time James Littman had performed his duties as shochet; that he was identified as the Reverend Littman, clergyman; and that Zamek was listed as a moneylender, might serve as some basis for mounting a counter-narrative about a case tinged with an antisemitic flavor, motivation, and intended result. The charges against Zamek underscore a reading of the events in Aberdeen in the summer and autumn of 1893 that goes beyond a simple summons on a charge of animal cruelty for one instance of things gone wrong in the slaughterhouse.

Zamek was merely present at the slaughterhouse on August 30. No doubt he was in attendance as president because this was the first occasion of shechita for the newly formed Jewish congregation of Aberdeen. He did not participate in the killing of the animal, nor did he play any role in the preparation of the bullock before Littman killed it. But he was charged as an active participant in the acts of animal cruelty by the Procurator Fiscal. No other person in attendance at the killing was placed in the dock, including the non-Jewish slaughterhouse employees who had prepared the animal and had, if the prosecution witnesses were to be believed, actively participated in casting the cow

34 "Aberdeen Liberal Association," *Aberdeen Journal*, March 19, 1879.

35 "Death of Mr. Robert Lamb, Advocate, Aberdeen," *Aberdeen Journal*, December 14, 1914; "Death of Aberdeen Procurator-Fiscal," *Scotsman*, December 14, 1914.

in a manner that was considered to be inherently cruel. Only the two Jews, one of whom had done nothing but look on, were accused of torturing the animal. The way in which the trial and verdict unfolded go some way to highlighting the meanings of law and legal principle, and anti-Judaism in the interpretation of anti-shechita cases, as they compete for dominance in the history of the criminal prosecution of the Jewish method.

The charges brought by the Procurator Fiscal related to three sub-sets of events in the killing of the bullock. The first charge dealt with the tying of the animal and casting it violently onto a bare concrete floor; the second maintaining it in a cruel position, with its head bowed for some period of time; and the third to the poor execution of the actual slitting of the beast's throat which was allegedly done in an ineffective manner, causing the bullock to thrash about and take several minutes, during which it was in pain, to die. The charges carry elements that go to the particular facts of the events of August 30, while others, such as the position of the bullock in preparation for the killing, and the act of killing the animal itself, are much more directly related to the standard practices of shechita. In each of these elements, they bear a striking resemblance to the allegations against Abraham Levitt twenty years later in Nova Scotia.

The first part of the case against Rabbi Littman and Alexander Zamek involved accusations relating to the cruelty of casting, the process of preparing the animal for the final stages before it is killed. The animal is cast so that it is kept still and its throat exposed for the single cut of the shochet's blade. At this time, casting involved different techniques according to local practice, but it usually entailed the use of ropes, rings, and sometimes pulleys to ensure that the animal was placed in the appropriate posture. At the same time, Jewish law requires that the process of casting not injure the animal since such an injury would make it unsuitable for slaughter for consumption by Jews, and Jewish ethics demand that the animal be treated with kindness and mercy throughout. Concerns about casting were abroad at the time and members of the Jewish communities themselves were at work to create a mechanism for casting that would cause minimal suffering to the animals and be consistent with the religious requirements for slaughter.[36]

The issue of casting would continue to arise from time to time in relation to the practice of shechita. Jewish communities were forced to justify the

---

36 "The Slaughtering of Animals," *Manchester Guardian*, August 29, 1893; "Editorial," *JC*, October 20, 1893.

humane nature of their mode of killing, including the preparatory stages.[37] New and improved technologies for immobilizing the animal with a minimum of distress would be introduced sporadically, the most well-known of which became the Weinberg casting pen, a device that turned the animal upside down, into a position in which its throat could be cut, without the use of the typical ropes and chains.[38] Casting continues to be an issue at the heart of attacks on the Jewish method to this day.[39] The fundamental ethical norms of Jewish religious practice demand that animals be treated with kindness and mercy, and the requirements of shechita impose recurring conflicts within some Jewish communities over the technical aspects of casting.[40] The ongoing emphasis on casting in prosecutions of shochetim, and the way in which allegations around the practice are made one with killing without stunning, reflect, in part at least, a fundamental and basic opposition to shechita as a matter of principle by many animal protection authorities. Such charges in relation to casting during criminal prosecutions for animal cruelty against the Jewish method are almost always thinly disguised attacks on shechita more generally.

In 1904, when another public campaign against the Jewish method of slaughter was under way, the Shechita Committee of the Board of Deputies of British Jews met on more than one occasion to discuss issues and concerns around casting. They heard reports of more failed mechanical inventions for the casting of cattle, and of the inadequate space and sanitary norms at the Deptford slaughterhouse. They learned that conditions at the Birkenhead lairages and the practice of casting there was far superior to that practiced in and around London. The Shechita Committee was perfectly aware of the concerns and issues surrounding the treatment of animals during this pre-killing phase. But for them, the difficulties arose not because of poor practices

37 "The Jewish Method of Slaughter," Letter from Charles H.L. Emanuel, Solicitor and Secretary of the British Board of Deputies, *Manchester Guardian*, September 13, 1924.

38 "Jewish Slaughter of Animals: A Leeds Invention," *Manchester Guardian,* December 6, 1928; "Humane Slaughter of Beasts: Manchester Trial of Casting-Pen," December 8, 1928; "Jewish Humane Slaughter: The Weinberg Pen," February 3, 1931.

39 Temple Grandin, "Improving religious slaughter practices in the U.S.," *Anthropology of food* [Online], 5 May 2006, Online since 01 May 2006, http://aof.revues.org/93 2006; Temple Grandin and Joe M. Regenstein, "Religious Slaughter and animal welfare: a discussion for meat scientists," *Meat Focus International* 1994; Temple Grandin, "Problems with kosher slaughter," *International Journal for the Study of Animal Problems* 1 (1980).

40 Roger Horowitz, *Kosher USA: How Coke Became Kosher and Other Tales of Modern Food* (New York: Columbia University Press, 2016), 224–243.

at Deptford. They flowed directly from the division of labor during shechita in English slaughterhouses.

Sir Simon Montagu reported to his colleagues on the committee that "the casting was done by non Jews [*sic*] & that any attempt to introduce Jewish casters would be strongly resented by the Christian casters; that there was a great question whether it would be advisable to substitute a Jewish slaughter-house where the whole operation would be under the supervision of Jews."[41]

The committee later resolved to propose the introduction of all Jewish operations where casting would be under the control and supervision of the shochet so that Mosaic law could be followed more strictly in the entire killing process.[42] The Shechita Committee of the Board of Deputies had a diametrically opposed view to the problems of casting, and their source, than did the prosecuting authorities and the inspectors of the RSPCA. They saw the lack of Jewish control over the process as being at the core of the ongoing controversies. Prosecutors and inspectors continued to act as though the shochet exercised control over all the preliminary aspects of the slaughtering of cattle under the Jewish method. In the Aberdeen case, Zamek and Rabbi Littman were charged, as was Rabbi Levitt in the Canadian case, with acts of cruelty, parts of which related directly to the process of casting. No Christian butcher or slaughter man, present in each instance, was charged. Jews complained of a lack of control, while prosecutors insisted on a narrative about the mode of slaughter as one carried out under a Jewish monopoly of power.

No matter what steps were taken throughout this period to ensure that cattle were humanely slaughtered, as commanded by Jewish law, prosecutions continued. While there is no doubt, especially given concerns in the Jewish community itself, that casting sometimes posed difficult issues about the treatment of animals, the persistence in attacking shechita as a whole can leave little room for arguments that the prosecutions were only based on the individual facts of particular cases. The unfolding of the Aberdeen attack on shechita, and the bringing of charges against the two Jewish officials present on the day, and no one else, demonstrate the actual normative marginality of concerns about casting in and of itself in the minds of those pursuing Zamek and Rabbi Littman.

---

41 Board of Deputies of British Jews, Shechita Committee, 3121/C16/1, November 17, 1904; November 16, 1904, London Metropolitan Archives (LMA).

42 Shechita Committee, January 30, 1905.

The prosecution in Aberdeen set out its list of five witnesses and the defense named twelve people to testify in behalf of the accused. John Campbell, a butcher of seven years' experience, and who was present on August 30, testified about the process of binding the animal and killing it. Although he was called as witness for the prosecution, Campbell was of little help on some of the charges. He testified that the bullock was put on the floor without much force and that Zamek was simply standing outside the door, and did not participate by giving instructions, while the preparations and killing took place. He declared that the animal took four minutes to die after its throat was cut, but under cross-examination he stated that he had seen animals kick as much and for as long when killed using the normal method involving stunning.

George Sinclair, who helped dress the bullock after it had been killed, testified that the animal had been lowered gently to the ground to be placed in a position to be killed and that "there was just the usual struggling." Charles Farquharson, butcher, said that he had spoken to Zamek to protest as a matter of principle against killing the bullock without stunning. After Farquharson testified about the animal being tied with its head down, the prosecution recalled Campbell to confirm that part of the evidence. Throughout the case, the Procurator Fiscal, supported by occasional interventions by the baillie, contended that maintaining the animal in this position was itself evidence of cruel and inhumane treatment. Witnesses for the defense simply stated that the position in which the animal stayed for a relatively short time before being lowered gently to the floor, was exactly the same stance the bullock would itself have adopted when eating. Such a natural position could in no way be considered cruel.

Veterinary surgeon James Thomson testified that preparing an animal for slaughter by tying it down to a ring and killing it without prior stunning was not a humane way of slaughtering. Thomson was carefully cross-examined on both the process of killing the bullock in the Jewish method and the stunning method, including whether he was aware of cases where more than one blow had been required to stun the animal and whether in his scientific opinion this would cause the animal to suffer. He was asked if the kicking of the animal following the cutting of its throat by the shochet would be indicative of the animal suffering pain. Thomson responded that the blow to the animal in killing using stunning caused pain and that the kicking following the slitting of the throat in the manner prescribed by the rules of shechita was caused by "nervous irritability" and was in no way a sign that the animal was suffering.

The defense strategy was clear and required by the nature of the legal norms and the rules of a criminal trial. Witness testimony was used to demonstrate that in this particular case the animal was not subjected to any sort of extraordinary treatment that would be seen to be exotic or outside the normal way in which animals were brought to slaughter. The only question of uniqueness was the absence of stunning and the defense sought to establish that stunning itself caused pain to the animal and often was not the efficient single blow its proponents described. Killing in the Jewish mode did not cause the animal to suffer beyond the initial quick cut with the *hallaf*. All other signs of cruelty such as movement of the limbs after the throat had been slit, were identical in both modes of slaughter and in no way indicative of pain since the nerve pathways to and from the brain had been completely severed. At a minimum, shechita was as humane as the stunning method and perhaps in a stronger version, it was quicker and more sympathetic to animal suffering and injury.[43]

The veterinary specialist called as the first witness for the defense, Dr. Garden, confirmed that the pain suffered from the shochet's cut would be the same as that endured from the stunning blow and that once the cut had been administered, there would be no further sensation of pain. He stated, on cross-examination by the Procurator Fiscal, that the position in which the animal had been originally kept, with its head bowed, was precisely the grazing position and likewise would not in and of itself have caused any pain. When pressed by the prosecutor, he did confirm that the animal, tied in such a position, might have experienced fear, but not beyond the fear suffered in the ordinary course of non-Jewish slaughter.

Jacob Feurst, a rabbi from Edinburgh testified in the second volley of the defense. He had carefully examined James Littman's qualifications. His training and appointment as shochet had originally come from the Chief Rabbi in Poland, "one of the most eminent men we have." His capacities as a shochet had been confirmed by British Jewish authorities in London, but on cross-examination he admitted that he was testifying as to Rabbi Littman's qualifications and experience as certified by ecclesiastical authorities in Poland and the United Kingdom, but had never witnessed James Littman slaughtering cattle. On re-examination, Rabbi Feurst asserted that Jewish customs condemned cruelty to animals. This brought the case back to the strict legal framework.

> The Baillie—The Jewish customs have nothing in the world to do with this.

---

43 This continues to be the position. Rabbi Solomon David Sassoon, *A Critical Study of Electrical Stunning and the Jewish Method of Slaughter (Shechita)* (Letchford, Herts: published by the author, 1955).

The defense returned to the legal and factual strategies that had to inform their evidence. A. Duff Dunbar, another veterinary surgeon, testified that the Jewish method was quicker and more painless than killing with prior stunning and that any signs of struggle after the incision would be attributable solely to "involuntary nervous power." The defense had led evidence, and elicited information from some of the prosecution's own witnesses, that Alexander Zamek had not participated in the preparations for the killing of the bullock or the slaughter itself. It had demonstrated that the animal has not been thrown to the floor in a violent or cruel manner as alleged, but on the contrary that it had been lowered slowly and gently. It had presented proof that James Littman was a highly qualified shochet and that as a general matter the Jewish method of killing was at least as humane as the so-called normal method of killing with prior stunning.

The next witnesses for the defense, Alex Keith and Archibald McKay, were fleshers who had witnessed the killing of the bullock on August 30, 1893. Each testified that the cutting of the throat had been quickly and skillfully done. Another flesher who had been present, John George, stated that he did not believe that the animal had suffered from standing with its head attached to the ring and bowed, or that there was anything out of the ordinary in the way the animal was treated. Two butchers, William Guthrie and Alexander Ross, affirmed the skillful nature of the killing carried out by the shochet. During the evidence offered by the defense, the Procurator Fiscal admitted. "I have not proved that it was put down roughly."

The prosecution lawyer himself removed the first alleged act of cruelty from consideration. He confirmed this in his summation to the Bench. He did however unmask the real legal position of the case in his further summation. He claimed "the regular mode of slaughter rendered the animal unconscious at the first blow, and that any other mode was exceptional." "The only question" before the baillie "was whether it was humanely slaughtered." While he was at pains to assert that the case had "nothing whatsoever to do here with regard to the religious persuasion of these gentleman," he went on to base his argument on the assertion that killing an animal without prior stunning was inherently cruel. The defendants were free to choose to kill an animal in any way they wished, as long as it was done "in such a way as to save it unnecessary pain." Since it was the prosecution's argument that killing the bullock without stunning always and by definition caused it unnecessary pain, the carefully worded argument that the case had nothing to do with the defendants' Judaism was rendered problematic in the extreme. If killing without stunning was by definition in violation of the statutory prohibition against animal cruelty, then the defendants were not in fact or in law free to choose their method of killing, and the case had everything to do with their religious beliefs. The choice facing the defendants under the prosecution theory of the case and its jurisprudential stance on the facts, was between obeying a religious command and living according to their faith, or complying with a law that declared their religious beliefs and practices to be a criminal offense punishable by up to three months in prison.

The defense submitted there was absolutely no evidence against Alexander Zamek, who like the non-Jewish fleshers and butchers who had testified in the case, and who had not been charged, had merely been present as an observer. As far as the case against Rabbi Littman was concerned, the defense solicitor argued that there was no evidence of cruelty in the way the animal had been treated before it was slaughtered, just as there was no proof offered by the prosecution that any ill-treatment of the bullock had occurred during the killing itself. "Mr. Littman was bound by the Jewish law to take the greatest possible care." This was as far as the defense would or could go in discussing the religion of the defendants. The argument presented was based on the facts and on the position that the prosecution had not made out its case. There was no evidence as admitted by the Procurator Fiscal that the animal had been thrown violently to the floor. Experienced witnesses had testified both that the animal was not made uncomfortable beyond the ordinary course of events and that the actual killing had been swiftly and competently carried out.

> Baillie Kemp—In view of the importance of the case I have taken very careful notes, and I purpose to consider the matter until Monday morning.

Animal cruelty prosecutions were not uncommon at the time. Despite the declarations both of the baillie and the Procurator Fiscal in his summing up, something more than a factual inquiry into an individual instance of the slaughtering of a bullock on August 30, 1893 in Aberdeen was at stake. Any number of individuals had been present at the killing of the animal. In addition to the shochet and the president of the Jewish congregation, numerous butchers and fleshers, as well as an inspector of the Society for the Prevention of Cruelty to Animals, had been in attendance. This was the first instance of shechita in the city and it had resulted in a large crowd to witness the event, and had culminated in a prosecution for animal cruelty. The consequences of adopting the approach that asserted that killing without stunning was in and of itself an act of animal cruelty contrary to the law would have had the effect of making the

Jewish mode of slaughter, whenever practiced, a criminal offense. The baillie understood this, and his declarations that religion had nothing to do with the case before him notwithstanding, he needed the weekend to grapple with the case and its implications.

The judgment that followed the recess offered more questions than answers.[44] Alexander Zamek was found not guilty, but in so far as Rabbi Littman was concerned, the baillie entered the unique Scottish verdict of not proven. For many, the verdict strikes an immediate discordant note. If the prosecution has not made out its case, has failed to meet its burden of proving all of the elements of the offense, then the defendant is not guilty.[45] In this case in particular, the idea that the not proven verdict lies as a taint on the defendant who appears to have managed to get away with something is particularly suggestive. Rabbi Littman was not guilty of animal cruelty, nor was he innocent of the charge. While there was a clear recognition that Alexander Zamek had done nothing, this did little to remove the fact that he had been charged with the animal cruelty offense and had stood trial on those accusations, merely because he was the president of the Jewish congregation. No non-Jewish butcher or assistant who had actually helped in casting the animal had been charged with any offense. The prosecution proceeded against the two Jews present because this was a frontal attack by the state against the Jewish method of slaughter. That Zamek was charged despite there being no evidence against him goes some way to allowing us to understand that the Aberdeen shechita case can be characterized as a direct and unmitigated attack by the state on the beliefs and practices of Orthodox Judaism. The Baillie persisted:

> In dealing with this case I think it proper to remark that I have not regarded the trial as an inquiry to determine whether the Jewish mode of slaughtering animals, as opposed to the method usually followed here, is or is not the better or more humane method of the two, or the reverse. I make this observation because frequently, while evidence was being led, questions were put by the accused's agent strongly suggestive of that being the matter under consideration, and the point to be determined.[46]

44 "The Hebrew Cattle Slaughter Case: Judgment Given," *Aberdeen Journal*, October 17, 1893. Quotations in this section come from this report.

45 A.S. Pringle, "The Verdict of 'Not Proven' in Scotland", 4 *Canadian Law Review* 248 (1905).

46 "The Hebrew Cattle Slaughter Case: Judgment Given."

This was nothing more than a simple charge of animal cruelty brought by the Procurator Fiscal under the relevant statute relating to the particular facts of the case. Other facts, for example that this was a charge brought immediately following the first killing by shechita in Aberdeen; that the Procurator Fiscal had sought to prove the charge of ill-treatment by establishing that the usual method of stunning was not cruel; that no non-Jewish participant in the killing in question had been charged; or that no prosecution appears to have been brought under the statute for animal cruelty when an animal had been killed in the usual way, were of no moment in allowing him to declare that this was a simple case in which religion played no role.

The magistrate recognized that there had been no evidence to indicate that the animal had been thrown violently to the floor and he dismissed that part of the case without further comment. On the question of tying the bullock to the ring with its head bowed for some time, the baillie was more assertive, qualifying the defense evidence that this was not cruel and simply replicated the grazing position as "too absurd to be taken seriously." The baillie was bound by Scots law on the question of the elements of the offense of animal cruelty. Proof of wanton and purposeless cruelty, or pain out of all proportion, was required in order to sustain a conviction.[47] The prosecution had not succeeded in making out a contravention of the statute, but again without further explanation, he had entered a verdict of not proven.

The defense moved for an award of costs in the case, given that Alexander Zamek had clearly been charged without reason, and that the case against Rabbi Littman had fallen short. The defendants were denied their costs, the verdict of not proven was left untouched because there was no appeal in the case, and there appears to be no further Scottish prosecution in which shechita was targeted as an instance of animal cruelty.

In January 1894 the Lord Provost (a position similar to that of Mayor, but carrying with it in the cities of Aberdeen, Dundee, Edinburgh, and Glasgow the function of lord-lieutenant) of Dundee, where anti-shechita agitation had preceded the Aberdeen prosecution, spoke at the annual meeting of the Society for the Prevention of Cruelty to Animals and specifically addressed the Jewish method of slaughter, which he declared to be according to the best available scientific evidence, at least as humane as the method involving stunning. Another speaker followed the Lord Provost and further informed his audience

47 *Todrick v Wilson* (1891) 18 R 41; 28 SLR 724. A different, lesser burden was imposed on the prosecution in English law, *Ford v Wiley*, 23 QBD 203 (1889).

that the worst possible way of dealing with any question of alleged cruelty in the practice of shechita, "was in a court of law" because then "it almost became a case of persecution."[48] Leaders of the Scottish community, and it would appear, the SPCA, had learned, perhaps belatedly, the hard lessons and pernicious consequences of the Aberdeen case.

On January 23, 1894, the first animal after the trial was killed by the Jewish method in Aberdeen, at the same premises in Hutcheon Street, again by Rabbi Littman.[49] The slaughter occurred in the presence of two inspectors from the Society for the Prevention of Cruelty to Animals, Alexander Zamek, and two local vets. "No hitch occurred in the process, which lasted about eight minutes."[50] The presence of two veterinary surgeons and representatives of the Society for the Prevention of Cruelty to Animals had nothing to do with the Jewish method, or the principle that stunning was the only method of humane slaughter.

The Subtexts of the Aberdeen Shechita Case

Having kept its editorial powder dry while the case was before the courts, the *Aberdeen Journal* made its feelings about the prosecution of the synagogue president and the rabbi and shochet clear once the verdict had been handed down.[51] The first sentence of its editorial characterized the incident as the "frivolous and vexatious Jewish slaughter case" indicating not just its general feelings, but that as far as the local paper was concerned, this was, and had always been an anti-Jewish case. The paper speculated that there must be some kind of "unrevealed history" at work, since its reading of the evidence adduced during the "libel," a word it repeated throughout its attack on the prosecution, was to the effect that if there were cruelty in animal slaughter it occurred every day in the ordinary operations of slaughterhouses throughout the land. For the *Journal*, the truth behind the prosecution was that the "nativity of the libel must be traced elsewhere than in Aberdeen" since no convincing evidence had been adduced at the trial. External machinations could be the only plausible explanation for the fact that despite taking an entire weekend to reach his decision in the case, Baillie Kemp had been unable to articulate anything beyond "the shabby stigma" of the not proven verdict.

48 "The Shechita Question in Scotland," *American Hebrew* (New York), January 19, 1894.
49 "The Jewish Slaughter of Cattle," *Aberdeen Journal*, January 24, 1894.
50 Ibid.
51 "Editorial," October 17, 1893. The discussion is based on the editorial.

The *Journal* wondered at the absence from the witness box of the SPCA informant who had told the Procurator Fiscal that the animal had been thrown violently to the ground. This absence was made more puzzling given that all the witnesses present had testified that the animal was lowered gently to the ground, and the prosecutor had been compelled to admit during pleadings that this part of the accusation was without foundation. The *Journal* called into question not just the factual basis of the prosecution, but the obscure role of the Society for the Prevention of Cruelty to Animals in mounting the whole "foolish and groundless" case in Aberdeen. When the Aberdeen branch of the SPCA met in April of the next year, it agreed that the case had revealed a prima facie case of cruelty, and that this justified "the inspectors and the secretary in bringing the matter under notice."[52] For the Aberdeen newspaper's editorial writer, "The Baillie was —perhaps superfluously—anxious to have it understood that his decision was all out of his own head; it would be more interesting to know in what head the accusation was conceived, and what good purpose such a prosecution could have been designed or expected to serve."

A leading voice of public opinion in Aberdeen, taking into account the laws of libel, made it as clear as it could, that there was more to the case than met the eye. The accusation was groundless; the evidence all fell in favor of the accused; and the verdict of not proven remained an ongoing taint on Rabbi Littman and the Jewish practice of shechita. The *Journal* contained a report the following week of a public lecture in which a leading member of the community had stated that he could not see the justice of the prosecution of the Aberdeen case and this was met with applause from those in attendance.[53] Whatever antisemitic motivation might be revealed by the decision to bring the case in the first place, and in the conduct of the trial and its aftermath in the juridical arena, many in Aberdeen were troubled by the clear libel inherent in this pursuit of their Jewish fellow citizens.

If the case had such implications for many Aberdonians, it was equally clear that the prosecution, or as the *JC* characterized it from the beginning, the persecution, of James Littman and Alexander Zamek, in Aberdeen was significant for the Jews of Britain as well. From the largest centers, especially London, Jews supported the small Aberdeen congregation not just out of a broad sense of solidarity with their coreligionists, but because an attack on shechita

52 "News," *Aberdeen Journal*, April 4, 1894; "The Society for the Prevention of Cruelty to Animals," March 31, 1894.

53 "Lecture by Sir Arthur Grant at Dyce," October 28, 1893.

in Aberdeen could quickly have turned into a United Kingdom wide assault on the Jewish mode of killing. From its earliest coverage of the case, the *JC* had been clear that the "whole question of Shechita is called into question."[54] Otherwise, the charges against Alexander Zamek, who had merely been present at the abattoir, were entirely inexplicable. The *JC* reported on the activities of the Aberdeen branch of the Society for the Prevention of Cruelty to Animals, and made clear both its belief that the SPCA was behind the prosecution and that there was little credence to the Society's protestations that the case was a singular instance of alleged animal cruelty and had nothing to do with an attack on the Jewish method of slaughter "provided always proper arrangements are made for carrying it out in as humane a manner as possible."[55]

The case was experienced in the broader social and political contexts of the United Kingdom as something more significant than an isolated animal cruelty prosecution based on, as the baillie and Procurator Fiscal claimed, the limited facts and the isolated acts of suffering allegedly inflicted on a bullock in a slaughterhouse in Aberdeen. The case and the verdict received extensive coverage in newspapers throughout Scotland and England.[56] The press received considerable correspondence on the subject of shechita, with predominance given in many cases to opponents of the Jewish method.[57]

The *Aberdeen Journal* had emphasized the scientific support for the practice of shechita, both in terms of reducing the suffering caused to the animal, and emerging medical evidence that kashrut generally, and shechita in particular, were more careful to impose sanitary demands that made the consumption of kosher food more healthy and safe for consumers.[58] There was significant support at this time, and later, among medical and veterinary professionals for arguments relating to the relative painlessness of the Jewish method,[59] and for

54 "The Shechita Question," October 6, 1893.

55 "The Shechita Question," October 13, 1893.

56 "Singular Prosecution of Jewish Ministers," *Northeastern Daily Gazette* (Middlesborough), October 4, 1893; "Jewish Mode of Killing Cattle," *Dundee Courier and Argus*, October 6, 1893; "Law and Crime," *Bristol Mercury*, October 7, 1893; "Jewish Slaughtering of Cattle at Aberdeen," *Yorkshire Herald and York Times*, October 17, 1893; October 21, 1893; "The Jewish Method of Cattle Slaughter," *The Times*, October 17, 1893.

57 "Jewish Cruelty to Cattle," *Hampshire Advertiser* (Southampton), October 28, 1893; "Jewish Cruelty to Animals," *Glasgow Herald*, October 18, 1893; "The Slaughtering of Animals," *Manchester Guardian*, September 26, 1893.

58 "Editorial," October 17, 1893.

59 "Tuberculosis Meat Dangers: The Jewish Method of Slaughter," *Guardian*, August 29, 1906; "The Slaughtering of Animals," *British Medical Journal* 1, June 9, 1894; "The Jewish Shechita," *British Medical Journal*, 2, October 21, 1893; "Anaesthetisation of Cattle Before Slaughter," *British Medical Journal* 1, May 31, 1884.

the broader sanitary health claims.[60] Much of the positive response came as a result of a concerted campaign by Jewish doctors and professionals, with the active involvement of Hermann Adler, the Chief Rabbi.[61] The contrary view on both issues, not without clear antisemitic undertones, competed in the professional literature.[62] Shortly before the events in Aberdeen, one prominent member of the veterinary world, Professor John McFadyean, had publicly described the Jewish method of slaughter as "revolting."[63] The Jewish campaign to present shechita in a positive light by emphasizing in equal measure the scientific aspects and theological sources met with some success since it combined appeals to the world of science in Victorian England, with direct invocations of divine authority that resonated with many Christian discourses abroad at the same time.[64]

In the United Kingdom, there was still ongoing debate and controversy about the humane nature of shechita before, during, and after the Aberdeen case.[65] It is little wonder that the prosecution of Alexander Zamek and James Littman attracted such attention throughout the country, or that evidence of antisemitic agitation caused British Jewry to take a collective interest in the outcome of the case and join, where possible, in the battle to prevent the broad criminalization of the Jewish mode of slaughter. In the mind of everyone but Baillie Kemp, the Procurator Fiscal, and the Aberdeen branch of the Society for the Prevention of Cruelty to Animals, the case against Alexander Zamek and Rabbi Littman was something more than a singular, and entirely secular, legal action.

Press coverage and commentary were quick to identify the international legal, social, and political contexts in which the Aberdeen prosecution took place. Various journals drew their readers' attention to the most direct

---

60 Hermann Adler, "Sanitation as Taught by the Mosaic Law," *British Medical Journal* 2, November 11, 1893; "The Sanitation of Mosaic Law," *Manchester Guardian*, September 25, 1899; "Bloodless Fish and Flesh," *Morning Post* (London), May 13, 1889; Mitchell B. Hart, *The Healthy Jew* (Cambridge: Cambridge University Press, 2007).

61 Israel Finestein, *Anglo-Jewry in Changing Times* (London: Vallentine Mitchell, 1999), 230–231.

62 "The Jewish Method of Slaughtering," *British Medical Journal*, 1, February 6, 1897: 345–346.

63 "National Veterinary Association," *Manchester Guardian*, July 27, 1893. Professor John McFadyean is described as "the first modern veterinary scientist in England." He joined the Royal Veterinary College as professor of pathology and bacteriology in 1891 and established a research institute in animal pathology at the College. http://www.rvc.ac.uk/about/the-rvc/history.

64 "Sanitation as Taught by the Mosaic Law." Finestein, *Anglo-Jewry in Changing Times*, 230.

65 "Public Slaughterhouses," *Manchester Guardian*, January 29, 1892.

and obvious point of comparison, the recent decision to ban killing without stunning in Switzerland, and the anti-Jewish attitudes that had informed the campaign to institute the ban. The *Aberdeen Journal* highlighted the fact that the charges against Alexander Zamek and Rabbi Littman "took the exact shape in which it has of late been brought on the Continent by anti-Semetic [*sic*] factions that have been intolerantly active."[66] The *JC* special correspondent in the earliest dispatches from northeast Scotland had highlighted the clear connections between the Aberdeen persecution and events in Switzerland.[67]

The invocation of the Swiss example is telling. It situates the ideas and ideology, and the knowledge, that so evidently circulated in the international circles of the anti-animal cruelty world. Twenty years after the Aberdeen case, in Nova Scotia, Robert Murray did not hesitate to invoke the German and Swiss examples following the decision of Magistrate Fielding criminalizing shechita, as Societies for the Prevention of Cruelty across North America clamored for the latest news and information from Halifax about his great victory for the helpless and dumb creatures of God.

The Swiss case was of much more temporal currency and perhaps of deeper significance in Aberdeen. A few weeks before the Scottish trial, the Swiss had banned the Jewish method of slaughter through a constitutional amendment by referendum. This was the first such use of the amending mechanism and while it has been argued that the shechita ban "does not hold an illustrious place in the annals of Swiss democracy," it does occupy a significant position in the history of European antisemitism.[68] The Swiss case, like that in Germany, is notable for the clear antisemitic motifs and motivations that had informed the politics and law of the animal welfare movement. It is significant in so far as it demonstrates, as did the Halifax case, the ways in which these attacks on Jews and Jewish religious practice were part of an international anti-cruelty movement, at the heart of which sat unambiguous antisemitic rhetoric and beliefs. As Mesmer points out, the Swiss case is particularly indicative of the antisemitism/anti-cruelty nexus because, unlike Germany, there were no dedicated antisemitic political parties in that country. In Switzerland, "the animal welfare organisations took on this role."[69] At the time of the Aberdeen trial, both proponents of the ban on the Jewish method, and defenders of religious freedom, were acutely aware

---

66 "Editorial," October 17, 1893.

67 "The Shechita Question," October 6, 1893.

68 Beatrix Mesmer, "The Banning of Ritual Slaughter in Switzerland," *Leo Baeck Institute Year Book* 52 (2007): 190.

69 Ibid., 189.

of the significance of the Scottish case for the advancement of their respective causes.[70]

In both the Halifax and Aberdeen shechita cases, the German and the Swiss examples of legal abolition and prohibition were invoked as positive instances of national and regional attacks on animal cruelty. They were deployed as positive examples to be followed by those in Nova Scotia and Scotland who sought to protect animals from the innate cruelty of shechita. As Efron has pointed out, and as the Halifax and Aberdeen cases and the rhetorical devices surrounding them demonstrate, the anthropological, theological, scientific, and political aspects of the anti-cruelty movement in general, and specific attacks on shechita always carried within themselves a distinguishing antisemitism.[71] Efron further highlights the fundamental difficulty facing German and Swiss Jews in relation to the campaign on the "ritual slaughter question" (*Schächtfrage*). There are insights into, and a direct parallel with, the defense raised against legal charges involving Jewish officials in both Halifax and Aberdeen, and in arguments presented in other cases. The legal semiotics of animal cruelty legislation and the operation of the criminal justice system combined to direct the discourse of anti-shechita cases in a particular direction.

The politics of the debate created a strange synergy between theology and science. Jews had to engage with attacks on theological aspects of their belief and practice by demonstrating scientifically that shechita was not cruel.

> How religious belief proved and in some ways required, or at least was able to accommodate scientific verification, was a measure of the all-encompassing nature of the debate itself. ...
>
> Orthodox Jews were not entirely untroubled by this development, but their immediate concerns when confronted by the campaign to abolish ritual slaughter were political and not theological; in this political framework they believed they had to invoke the authority of science in staking out their claims. Both their responses and those of the non-Orthodox represented a novelty in the history of Jewish apologetics.[72]

---

70 "A curious question," *Times*, October 17, 1893; "Question of Cruelty to Animals," ibid., October 16, 1893; "A Question of Cruelty to Animals," September 11, 1893.

71 John M. Efron, "The Most Cruel Cut of All? The Campaign Against Jewish Ritual Slaughter in Fin-de-Siècle Switzerland and Germany," *Leo Baeck Institute Year Book* 52 (2007).

72 Ibid., 168.

What is highlighted by the invocation of the German and Swiss cases in the shechita trials in Scotland and in Nova Scotia is more than the international circulation of knowledge and rhetoric among various anti-cruelty advocates. In both instances, we find concrete manifestations of Efron's idea, but in the particular context of the limits of criminal animal cruelty prosecutions. The politics and science of anti-cruelty agitation and Jewish defenses of shechita required a scientific response of a particular kind because the nature of a penal trial on animal cruelty charges virtually compelled the shochetim to assert and to prove that an essential element of the offense was absent, i.e. that the Jewish method was at least as humane as the stunning method. The politics of anti-shechita agitation took on this juridified color in Britain and North America because of the peculiarities of the legal matrix at work at the time in any such cases. While there was significant scientific support for the humane nature of shechita, there was countervailing argument coming from veterinary authorities, medical practitioners, and with unwavering consistency from animal protection organizations. For every learned treatise by physicians such as Dembo, the key scientific text defending the Jewish method and proving the absence of suffering during shechita, there were professors from the University of London attacking the revolting cruelty of the Jewish method.[73]

Jewish lobbying in Britain and North America, combined with clear historical differences with the manifestations of antisemitic agitation in mainland Europe, meant that a blanket legislative or constitutional ban on shechita was not on the cards, despite the best efforts of Societies for the Prevention of Cruelty.[74] But the common law (and the criminal law of Scotland), combined with general provisions of statute law regulating and criminalizing cruelty to animals, meant that individual cases prosecuting shochetim could be strategically and tactically deployed to establish legal precedents the combined effect of which might have been (as was hoped to have been) the outright criminalization of the Jewish mode of slaughter. The pursuit of this prosecutorial tactic was not subject to direct political control or influence through considerations of a broader public interest in ensuring religious freedom or even in preventing the rhetoric in which antisemitism could become part of open

73 J. A. Dembo, *The Jewish Method of Slaughter* (London, 1894). The full title continues *Compared with Other Methods from the Humanitarian, Hygienic, and Economic Points of View.*

74 Report of the Committee Appointed by the Admiralty to Consider the Humane Slaughtering of Animals.

social, political, and legal discourse, since the legislative scheme created the public/private dynamic of enforcement of anti-cruelty laws by the Societies.

In Britain and the common-law jurisdictions of North America, the antisemitism of the anti-cruelty advocates and their campaigns against shechita could be more successfully elided through discourses of legal positivism, the practice of judicial neutrality, and other components of rule of law ideology, and the scientific neutrality of expert veterinary medical testimony on animal suffering. In the late nineteenth and early twentieth centuries, only a few years after Jewish emancipation, and long before any real concretization of a constitutional normativity of equality and religious freedom, Jewish defenders of shechita, and Jewish defendants in anti-shechita prosecutions such as that in Aberdeen in 1893, were compelled to confront legally and scientifically neutralized antisemitism in the guise of the prevention of cruelty on its own terms.

Outside Scotland, the *Times* sought to strike a somewhat more tempered approach to the case than had the *Aberdeen Journal*.[75] The paper's analysis would not have offered much solace to supporters of Jewish religious practice. After exploring the facts of the case and the findings of the baillie, the London newspaper reviewed the ongoing controversy in scientific opinion about slaughter without stunning and returned to its focus on the Swiss ban.[76] With these contextual matters in mind, it highlighted the simple fact that the nature of the Aberdeen case, like all other anti-shechita animal cruelty prosecutions, was specific and limited to the facts of the particular instance. The broader question remained open:

> It is still, however, open to contention that, in other cases, the preparation of 'Kosher' meat, in the manner prescribed by the extreme orthodox school, may be held to be a violation of the law. It is not altogether surprising that the attempt to put down these ritual practices by prosecutions has caused a good deal of alarm among Jews who adhere rigidly to rabbinical teaching....[77]

The *Times'* characterization of obedience to the laws of kashrut as extreme and its followers as rigid led almost naturally to the paper's broader conclusion on the issue of cruelty and slaughter. As is almost always the case in anti-shechita

75 "News," October 17, 1893.

76 "A Question of Cruelty to Animals," *Times*, October 16, 1893; "A Question of Cruelty to Animals: From a Jewish Correspondent," September 11, 1893.

77 "News," October 17, 1893.

discussions, antisemitism is specifically abjured and the ideals of liberal religious tolerance given voice, but within limits where "the principles of toleration enjoin that their scruples of conscience shall be respected if they conflict with no higher law." That higher law to which religious practice must adhere includes the humane concept of cruelty. Shechita is not to be criminalized or banned unless and until it can be shown to be cruel. "It is on the ground of cruelty, and on that alone, if cruelty can be established, that we should be justified in interfering by legal action with a religious scruple analogous to many others with which we should not think of meddling."

While the principles of religious tolerance might operate against the pursuit of shechita in the present circumstance, for the *Times,* the issue really comes down to issues of proof and scientific knowledge. If it could be shown by the weight of scientific evidence that killing without stunning was cruel, then humanity would trump extreme and rigid religious practices. The Aberdeen case demonstrated the lack of scientific consensus on the cruelty of killing without stunning the animal. It served to highlight that the daily killing of calves, sheep, and pigs was being undertaken without prior stunning throughout the United Kingdom. Targeting shechita alone "would seem to be straining at a gnat and swallowing a camel." The issue was not one of religious tolerance or liberty as a presumptive value of any significance. Instead religious freedom would exist only so long as science remained unclear and broad slaughtering and butchering processes of killing animals without stunning persisted. If and when those practices were eliminated, on the basis that science established that stunning was required to avoid cruelty, then shechita could unproblematically be banned. In such circumstances, shechita could be outlawed. It is little wonder that the Jews of Aberdeen, and Anglo-Scottish Jewry more widely, were troubled by the implications of the Aberdeen prosecution.

The two historians of Scottish Jewry who have addressed the Aberdeen shechita case have reached somewhat different conclusions about its underlying motivations. Collins arrives at his view on this question because of the historical origins of the anti-cruelty movement and its overt connections with an evangelical Christian ideology.[78] While he offers a nuanced analysis and accepts that there may have been more than one motivating factor behind the prosecution, Collins does highlight the arguments put forward by the *Aberdeen Journal* about hidden local and national realities behind the decision to prosecute. He makes

78 Brian Harrison, "Religion and Recreation in Nineteenth-Century England," *Past and Present* 38, no. 1 (1967).

it clear that the consequences of a successful case in Aberdeen could easily have led to a United Kingdom wide ban on shechita.[79] On the other hand, Abrams expresses much less concern on this issue, declaring, "the case was probably brought because of genuine concern for the welfare of the animal."[80]At its annual meeting following the failed prosecution, the local SPCA specifically addressed and refuted accusations of antisemitism. "Now, they respected the Jews for their history and for many of their qualities of the present time, but the Society had to do their duty to Jew and Gentile alike. It was explained that the Society incurred no expense in connection with the Jewish slaughter case."[81]

Abrams underplays the contemporaneous views of the editor of the *Aberdeen Journal* and the careful study the local newspaper made of the case. It is true that there was a genuine concern for animal welfare among members of the SPCA. But the way in which the ordinary method of striking a brutal blow to the skull of an animal, a method that evidence showed was not always effective, was contrasted with the Jewish slaughter method, gives rise to a clear implication under which cruelty and Jewish practice become synonymous. When combined with the widespread interest in the Jewish slaughter case; the vicious attacks in letters to the editor throughout the UK following the failure of the prosecution; the absence of non-Jewish defendants who had played central roles in the killing at the Aberdeen slaughterhouse; and all of the other factors mentioned by the *Aberdeen Journal* editorial, the key role played by antisemitism in this and other anti-shechita cases cannot be subsumed or removed, no matter how much genuine concern for humane slaughter was expressed by anti-cruelty advocates.

The confirmation that the Society incurred no costs in relation to the case supports the idea that while the SPCA may have instigated the action against the shochet and the president of the congregation by sending its inspector to the slaughterhouse to witness the first killing by the Jewish method in Aberdeen, the case itself was carried out at the discretion of the public authorities in the person of the Procurator Fiscal. This adds to the significance of the Aberdeen shechita case because it becomes the only prosecution, among all the known British instances, by state authorities, and not by a private/public animal protection society, against a Jewish community's religious practices.

79 "The Aberdeen Shechita Case."

80 *Caledonian Jews*, 22.

81 "The Shechita Question," *JC*, April 6, 1894.

In support of Abrams' position, or as a modification, when designs for a new, ultramodern slaughterhouse for the city were presented to the Council, four and a half years after the trial, they included drawings for a separate building accommodating Jewish slaughtering facilities. Shechita, and the existence of a market for kosher Aberdeen beef, had become part of the local meat industry's economic outlook.[82]

Collins claims that "the Aberdeen shechita case proved to be unique in the English-speaking world."[83] This statement is belied by the prosecution of Abraham Levitt in Halifax twenty years after the Aberdeen case. Even if we were to limit Collins' conclusion to the second half of October 1893, and not take the later Halifax case into account, he would still not have been correct. The *Aberdeen Journal's* detailed account of the trial itself included defendants' counsel's concluding remarks to Baillie Kemp where he "quoted a case heard before the Lord Mayor of London, in the Mansion House, in October, 1855, in which a charge somewhat similar to the present had been dismissed."[84]

The Aberdeen case was not unique in the English-speaking world; it was not even unique in the United Kingdom.

82 "Proposed Public Slaughterhouse: Reports by Burgh Surveyor," *Aberdeen Journal*, March 30, 1898.

83 "The Aberdeen Shechita Case," 92.

84 "The Hebrew Cattle Slaughter Case," October 14, 1893.

CHAPTER 3

The Mansion House Case: Anti-Shechita in London, 1855

Animals, Law, and the Victorians

In the early autumn of 1855, the Royal Society for the Prevention of Cruelty to Animals brought charges against Thomas Reynolds for ill-treating an ox by tying its head "to the ground by an iron ring round the lower jaw attached to a staple."[1] The charge against Reynolds related directly to the further accusation against Yankoff Cohen, a shochet, who was pursued by the RSPCA for cruelty to the ox "by improperly cutting its throat."[2] The alleged impropriety related simply to the cutting of the throat without prior stunning, in other words, to the practice of shechita. Thus began what appears to have been the first recorded case in English law of animal cruelty charges against a shochet performing his religious duties in killing an animal for food following the laws of kashrut. In this first case, the non-Jewish participant was charged with cruelty in relation to the process of casting the animal.

The first British statute punishing cruelty to animals as a separate offense, aimed at protecting animals themselves rather than any rights or interests of the owner or possessor of the chattel, was [a]n Act to prevent the cruel and improper treatment of cattle, known as Martin's Act.[3] Cattle was a generic term referring to "Horses, Mares, Geldings, Mules, Asses, Cows, Heifers, Steers, Oxen, Sheep, and other Cattle."[4] The statute made it an offense to "wantonly and cruelly beat, abuse, or ill-treat" any of the named animals and called for a summons before a Justice of the Peace or a Magistrate, who could inflict a

1 "The Police Courts: Mansion House," *Daily News* (London), October 17, 1855.

2 Ibid.

3 III Geo. IV, c. 71 (1822).

4 Ibid., §1.

penalty of a fine between ten shillings and five pounds. Failure or inability to pay the fine would result in imprisonment for up to three months. Martin's Act was replaced in 1835 by the Cruelty to Animals Act.[5] This statute added "bulls, dogs, bears, goats and sheep" to the list of protected animals, and was itself replaced in 1849 by An Act for the more effectual prevention of cruelty to animals.[6] The 1849 Act amended, extended, and clarified the nature of the offense. It covered any persons who "cruelly beat, ill-treat, over-drive, abuse or torture" any animal or who procured the commission of these acts and imposed a fine of five pounds on offenders.[7] It clarified the prohibitions on bull, bear, or badger baiting, and banned dog and cock fighting.[8] The Act imposed regulations on premises where cattle were to be slaughtered and on the preparation of cattle for slaughter.[9] Charges would be heard before a Justice of the Peace or Magistrate.[10]

This statutory evolution owed much to the Society for the Prevention of Cruelty to Animals, to which the Royal prefix was attached under Queen Victoria.[11] By 1867, the annual meeting of the RSPCA heard that in the previous year, the Society had obtained 375 convictions for cruelty to animals in London alone, and a further 401 outside the capital.[12] Three key elements from the early institutional history must be mentioned in the context of anti-shechita actions. The SPCA was empowered to investigate and prosecute instances of animal cruelty. Its history is the history of the public/private distinction and its operation in Victorian England, and from there to other parts of the United Kingdom and the empire. The model was copied in the North American context in both Canada and the United States. A socio-legal study of the RSPCA could reveal important aspects of the operation of the distinction, and particularly of the place of the state, even in enforcing laws, at this time.[13]

---

5 An Act to consolidate and amend the several laws relating to the cruel and improper treatment of Animals, and the Mischiefs arising from the driving of cattle, and to make other provisions in regard thereto, 5 & 6 Will. 4, c. 59 (1835).

6 12 & 13 Vict., c. 92 (1849). This statute would be amended by a further Act in 1876. An Act to amend the Law relating to cruelty to Animals, 39 & 40 Vict., c. 77 (1876) regulating vivisection.

7 Ibid., § 1.

8 § 2.

9 §§ 6–9.

10 §§ 13 and 28.

11 Arthur W. Moss, *Valiant Crusade: The History of the Royal Society for the Prevention of Cruelty to Animals* (London: Cassell, 1961).

12 "Prevention of Cruelty to Animals," *Observer* (London), July 14, 1867.

13 James Turner, *Reckoning with the Beast* (Baltimore: Johns Hopkins University Press, 1981), 40.

It could highlight the normative values of humanity, morality, and religious progress that informed the legislative changes and the vigorous enforcement of those changes in the hands of the Society.[14]

Secondly, from its founding in the 1820s and throughout the period under review when it launched its prosecutorial attacks on shechita, the RSPCA adopted and concretized a particular set of values and a peculiar worldview that was inherently Christian in nature. While it was in its original phases more eclectic in its membership and even in its ideological construction of the ethical foundations of anti-cruelty, by the 1840s, the SPCA had expunged non-Christian ideas. Louis Gompertz, a Jew, and one of the most dynamic forces in the early years of the organization was effectively purged as being too radical. "During the high tide of this intolerance, all radical and unchristian ideas were chastised as harmful to the anti-cruelty movement, just as they were harmful to the state. The first major split within the movement in the early 1830s arose partly from this strong commitment to both political and religious orthodoxy."[15]

It is clear that the history of anti-shechita prosecutions brought by the RSPCA, and its equivalents in other jurisdictions, was always confronted and contested by theological claims within Judaism about the centrality of kindness and humanity in the Mosaic traditions. This is not to suggest that the RSPCA did not maintain relations with the English Jewish community or that English Jews were disinterested in animal welfare. The situation was much more complex, as RSPCA attitudes towards Jews reflected broader social opinion and class biases, including anti-Jewish ideas. For James Turner "only after cooperation had lost all taint of disreputability did the RSPCA once again publicly associate with them– and then only with the safest and most eminent Jews, men like Baron Rothschild and Rabbi Marks. To have behaved otherwise would have jeopardized the SPCA mission to the middle classes."[16]

In Victorian England, and with obvious exceptions elsewhere, anti-cruelty activism in the hands of the RSPCA was informed by a distinct ideology of Christian social progress, in turn underscored by a clearly understood

14 Harrison, "Religion and Recreation in Nineteenth-Century England."

15 Chien-hui Li, "A Union of Christianity, Humanity, and Philanthropy: The Christian Tradition and the Prevention of Cruelty to Animals in Nineteenth-Century England," *Society and Animals* 8, no. 1 (2000), 269; "Mobilizing Christianity in the Antivivisection Movement in Victorian Britain," *Journal of Animal Ethics* 2, no. 2 (2012).

16 Turner, *Reckoning with the Beast*, 58.

class bias.[17] The link between the humanitarian treatment of animals and class control has been noted in the actions of Victorian animal welfare organizations. The Society for the Prevention of Cruelty to Animals, perhaps the most famous of all such institutions, was founded in 1824 (and made Royal by the patronage of Queen Victoria in 1840). Animal welfare organizations demonstrated an obsession with restricting the pleasure and work habits of the urban poor, while broadly turning a blind eye to the cruel practices of the middle and upper classes.[18] These ideas were not necessarily simplistic, either in their overarching ideological structures or in their constructions of the moral bases for anti-cruelty activity.[19]Abject poverty and criminality (with prostitution providing an important and gendered link), alcoholism, and disease characterized much of urban life in Victorian Britain. Nor can one question the role played by Christian advocates of social improvement in helping the worse off, the destitute, and the poor.

The final element is the identifiable antisemitic effect of its campaigns. A primary focus of the Society was always to be found in its attempts to change the prevailing practices in slaughterhouses, both in terms of sanitation and hygiene, but more specifically in relation to the evolving implementation of humane slaughtering methods. As Moss explained,

> In the first hours of its work, the humane slaughter of food animals was put amongst the Society's primary objects....
>
> The Society had always felt that the basic principle for the slaughter of all animals was that there should be effectual stunning before blood was drawn.[20]

Whatever the intentions or motivations of those seeking to impose the Society's understandings of humane slaughter, a primary effect of insisting on pre-slaughter stunning was to bring the RSPCA throughout its history into direct conflict with Jewish communities who insisted that their religious beliefs

---

17 Hilda Kean, *Animal Rights: Political and Social Change in Britain since 1800* (London: Reaktion, 1998).

18 Sarah Amato, *Beastly Possessions: Animals in Victorian Consumer Culture* (Toronto: University of Toronto Press, 2015), 47.

19 Harriet Ritvo, "Animals in Nineteenth-Century Britain: Complicated Attitudes and Competing Categories," in *Animals and Human Society: Changing Perspectives*, ed. Aubrey Manning and James Serpell (New York: Routledge, 1994).

20 *Valiant Crusade*, 69.

compelled them to kill animals without stunning. The ideological and normative frame of RSPCA activists who advocated the centrality of stunning in animal slaughter for human consumption put them on an inevitable collision course with observant Jews. Christian humanity was soon opposed to a seemingly innate and inescapable Jewish cruelty in the minds of those for whom God's dumb animals were a priority.

Alongside the specific statutory regime relating to cruelty offenses, and the powers of RSPCA inspectors, driven by the Society's normative outlook, other legislative provisions filled in the social, political, and legal frame in which attempts to criminalize the Jewish mode of slaughter would occur. In London, where the first anti-shechita case would take place, the mid-nineteenth century was the time of the statutory consolidation of the Metropolitan Police, who were given significant powers to deal with a vast array of offenses in the city.[21] Local police courts were restructured and given a firmer legislative basis and remit.[22] Local authorities themselves had significant powers to implement by-laws regulating slaughterhouses in terms of sanitation, hygiene, and public health, and their authority extended to regulating conditions in these establishments, as well as knackers' yards, to prevent cruelty to animals destined for slaughter.[23]

The regulation of animal slaughter, on the grounds of hygiene, nuisance, and consumer welfare, led to this evolving system of local government control over markets and slaughterhouses, and at the behest of the RSPCA, to the abolition of private abattoirs.[24] As urbanization continued to affect the changing social geography of British cities, and the surrounding countryside, the historical placement of markets in city centers began to cause offense to newly delicate Victorian sensibilities.

> This was precisely the wrong kind of visual economy sought by urban reformers, evoking something uncomfortably organic, untamed, moist, and visceral, rather like the fair, which became problematic at around the same time. Smithfield, London's oldest market, had been the site of slaughter, fairs, executions, floggings, and bearbaiting for centuries. Slaughterhouses infected and barbarized those around them, mysteriously stimulating drinking, fighting, and prostitution.[25]

---

21 An Act for further improving the Police in and near the Metropolis, 2 & 3 Vict., c. 47 (1839).

22 An Act regulating the Police Courts in the Metropolis, 2 & 3 Vict., c. 71 (1839).

23 Towns Improvement Clauses Act, 10 & 11 Vict., c. 34 (1847), § 128.

24 Bye-Laws for Slaughterhouses, City of London (London, 1877).

25 Chris Otter, "Civilizing Slaughter: The Development of the British Public Abattoir, 1850–1910," in *Meat, Modernity, and the Rise of the Slaughterhouse*, ed. Paula Young

The proximity of these facilities to the burgeoning commercial center of the imperial economy, the City of London, caused obvious consternation, especially among those who saw the brutality of the market as the literal embodiment of the social decay and criminality that went hand in hand with animal cruelty.[26] The increasingly Jewish character of nearby Spitalfields Market only exacerbated the concerns of the good Christian gentlemen and ladies of the RSPCA, and served as visceral reminders of their preconceptions about Jewish brutality, inhumanity, and obsession with blood.[27] The City itself, as the home of financial capital, was the center of attention for antisemites devoted to theories of Jewish perfidy.[28]

By the middle of the nineteenth century, in London and throughout the United Kingdom, animal rights were protected by a significant and sophisticated system of governance. The Royal Society for the Protection of Animals provided a key public/private nexus of investigation and enforcement, as well as an ongoing and vigorous institutional setting for lobbying for legislative change. Police and court structures had been formalized and a prosecutorial and punitive state administrative structure established. Local governments saw to the day-to-day regulation and inspection of slaughterhouses. A relatively effective and efficient system for governing and ensuring animal welfare by preventing cruelty and abuse was now in place. Into that midst of modernized governmental and police structures came Orthodox Jewry with its ancient laws of animal slaughter.

The Jews of London 1855

The position of Jews in London in 1855 forms a key part of the background of both the law and politics of the first English legal attack on shechita. Jews had been expelled from England in 1290 and remained outside the country for over three hundred years. Following their gradual return in greater numbers in the late seventeenth and early eighteenth centuries, by the middle of the next century, there was a significant Jewish minority population

---

Lee (Hanover, NH: University Press of New England, 2008), 91; Ian MacLachlan, "Humanitarian Reform, Slaughter Technology, and Butcher Resistance in Nineteenth-Century Britain," ibid.

26 Robyn S. Metcalfe, *Meat, Commerce and the City: The London Food Market, 1800–1855* (London: Pickering and Chatto, 2012).

27 Geoffrey Alderman, *London Jewry and London Politics, 1889–1986* (London: Routledge, 1989), 38.

28 David Blaazar, "Sharks and Shylocks: Englishness and otherness in popular discourse on 'the City', 1870–1914," *National Identities* 18, no. 2 (2016).

throughout the country.[29] At the beginning of the 1850s, the 18,000–20,000 Jews of London represented the vast majority of the country's total Jewish population of 35,000.[30]

While there was now a Jewish presence in England, Jews still suffered from a number of civil and political disabilities. Doubts remained over the validity of Jewish marriages, for example, and Jews were barred from holding many public offices, because the form of oath for these positions required the potential candidate to swear on his faith as a Christian. Years of political and legal agitation eliminated these impediments, although change was controversial and painfully slow.[31] The Jews Relief Act of 1845 removed some barriers against Jewish office holders.[32] The Jewish Relief Act of 1858 extended the rights of Jews to proffer an oath that would not violate their religious faith.[33] In 1866, the Parliamentary Oaths Act removed the final obstacles to full Jewish participation in the political life of the United Kingdom.[34] At the same time, there were and are complexities of emancipation beyond formal legality in the lifeworld of British Jews. Emancipation in broader social and intellectual contexts is much more subtle and complex, not just in defining exactly what emancipation might mean, but in requiring us to situate it "as a process, out of a mosaic of changing social, economic, and religious circumstances over which the Anglo-Jewish community had relatively little control."[35] The struggles over shechita embody

---

29 H.S.Q. Henriques, *The Return of the Jews to England: Being a Chapter in the History of English Law* (London: Macmillan, 1905).

30 V.D. Lipman, "A Survey of Anglo-Jewry in 1851," *Jewish Historical Society of England–Transactions* 17 (1951–1952).

31 M.C.N. Salbstein, *The Emancipation of the Jews in Britain: The Question of the Admission of the Jews to Parliament, 1828–1860* (Rutherford, NJ: Fairleigh Dickinson University Press, 1982); H.S.Q. Henriques, *The Jews and the English Law* (Oxford: Oxford University Press, 1908).

32 An Act for the relief of persons of the Jewish religion elected to municipal offices, 8 & 9 Vict., c. 52.

33 An Act to provide for the relief of Her Majesty's subjects professing the Jewish religion, 21 & 22 Vict., c. 49 (1858).

34 An Act to amend the Law relating to Parliamentary Oaths, 29 Vict., c. 19, although it was not until 1871 that all statutory impediments were formally abrogated. An Act to repeal divers enactments relating to Oaths and Declarations which are not in force; and for other purposes connected therewith, 35 Vict., c. 48 (1871).

35 Geoffrey Alderman, "English Jews or Jews of the English Persuasion? Reflections on the Emancipation of Anglo-Jewry," in *Paths of Emancipation: Jews, States, and Citizenship*, eds. Pierre Birnbaum and Ira Katznelson (Princeton: Princeton University Press, 1995), 129; Jacob Katz, *Out of the Ghetto: The Social Background of Jewish Emancipation, 1770–1870* (Syracuse: Syracuse University Press, 1998).

the ways in which these complexities, and resistance to Jewish equality, played themselves out in anti-cruelty agitation and prosecutions.

When the London trial of a shochet for animal cruelty took place in 1855, Jews in the United Kingdom were still engaged in the difficult struggle for full and unqualified legal emancipation. The 1845 Jews Relief Act and other statutory changes had to some extent begun the formal process of enshrining legal guarantees of equality and religious freedom, but the battle was far from over. The second Relief Act was still three years away and the final removal of religious impediments to the oath of office would not occur for eleven years.[36] Attacks against shechita would have been experienced in this political, social, and legal context as a threat to hard won, yet still fragile, religious freedoms and as a potential roadblock to full emancipation and equality. A single prosecution before a local magistrate under a statute relating to cruelty and ill-treatment of animals went well beyond the simply parochial nature, and limited factual circumstances, of the formal legal case.

The institutional arrangements within the Jewish communities of London, particularly in relation to shechita and its governance, offer insights into the ideological struggles of the time, and their management by London Jewry. The practice of shechita was an example of the self-regulatory functions and capacities of Jews, even before their formal emancipation in the second half of the nineteenth century. The internal, private governance of the mode of slaughter runs in intriguing parallel with the privatization of the enforcement of animal cruelty legislation in the hands of the SPCA. The public battles over cruelty accusations and prosecutions of shochetim and other Jewish officials took place in the unique legal, normative, and theological matrix that blended the public and private on both sides.

Todd Edelman writes that "the most salient characteristic of English Jewish religious life in the eighteenth century was its laxity and ignorance."[37] While there were many who practiced a traditional Orthodox Judaism, including following the rules of kashrut, many were looser in their observance of dietary laws.[38] The influx of Jewish immigrants from eastern Europe, both in terms of numbers, and because of their more traditional versions of religiosity, would drive up demand for products, including meat, produced

---

36 Michael Clark, *Albion and Jerusalem: The Anglo-Jewish Community in the Post-Emancipation Era 1858–1887* (Oxford: Oxford University Press, 2009).

37 *The Jews in Britain, 1656 to 2000* (Berkeley: University of California Press, 2000), 54.

38 Ibid., 115.

slaughter, did not come into existence until 1892.[13] Whatever intra-Jewish disputes may have arisen within the complex sociology of the Jewish communities in Manchester, from the earliest days of the eighteenth-century establishment of a permanent Jewish settlement in the city, observance of the rules of shechita was a central element of religious life.

The Prosecution

When the case opened in the Manchester Police Court, it was more than a minor penal matter before the lower ranks of the local judiciary charged with the enforcement of the ordinary and mundane cases of the criminal justice system of large cities. This was another attempt by local RSPCA officials to bring a halt to the practice of shechita they considered inherently cruel and inhumane. The Manchester prosecution must be understood not simply as one of a set of otherwise isolated local cases attacking the Jewish method of slaughter. It must be placed both in the national context of an unrelenting campaign by the RSPCA to introduce compulsory stunning of all animals in English slaughterhouses, and in the context of the transfer of knowledge among humane societies, especially in Europe. Only four years before the Manchester case, London had hosted the international congress of Societies for the Prevention of Cruelty to Animals and associated organizations, as well as the international Jubilee meeting of the RSPCA at the Royal Albert Hall.[14] Representatives from the United States, Algeria, Belgium, Egypt, various German cities, India, Italy, Russia, and Scotland were in attendance. Painless killing in slaughterhouses was a central topic for those present.[15] "Resolutions urging on the Legislature the subject of cruelty in slaughtering animals, and detailing the methods most approved, while condemning the Jewish method were agreed to."[16]

The Manchester anti-shechita campaign by the RSPCA must be placed in these local, national, and international contexts, including the unchanging position of the RSPCA and its international colleagues against the Jewish method of slaughter.[17] The case was heard before the Chief Stipendiary

13 Bill Williams, *Jewish Manchester: An Illustrated History* (Derby: Breedon, 2008), 65–66.

14 "Prevention of Cruelty to Animals," *Times*, June 23, 1874.

15 "Prevention of Cruelty to Animals," *Times*, June 18, 1874.

16 "News," *Freeman's Journal* (Dublin), June 19, 1874.

17 Damien Baldin, "De L'Horreur du Sang à l'Insoutenable Souffrance Animale: Élaboration sociale des régimes de sensibilité à la mise à mort des animaux (19e–20e siècles), *Vingtième Siècle* 123 (2014); "Notes of the Week," *JC*, June 15, 1894, on the international conference in Berne where the Jewish method of slaughter was one of only two items for discussion.

according to Jewish law.[39] Other forms of immigration from Germany, for example, revealed fractures and disagreements within different elements of London Jewry. Modernizing elements within Reform Judaism sought to modify Jewish religious practice with a particular focus on kashrut.[40]

This is not to suggest that the Jewish community of London from its earliest days rejected the dietary laws. The returning Sephardic community had established governance structures for all religious matters and concerns, including by 1696 a monopoly over shechita and the supply of kosher meat, for which it received tax revenue.[41] The arrival of immigrants throughout the next two hundred years gave rise to jurisdictional disputes in matters of ritual and observance as Sephardic and Ashkenazi groups sought to establish their mode of worship, *minhag*, collective governance, and obedience to the laws of kashrut. This manifested itself in the field of shechita in two legal controversies. Different rabbis and different rabbinical groups granted authority to shochetim to perform slaughter according to Jewish law, leading to challenges between and among religious leaders and butchers, and the Jewish public around the conformity of meat with the theological requirements of kashrut.[42] Heated controversies arose among different groups within London's Jewish communities over the laws of shechita and the competence of the city's shochetim in the mid-eighteenth century between 1755 and 1766.[43]

From as early as 1696, and throughout the eighteenth, nineteenth, and into the twentieth centuries, different shochetim brought libel actions against various shechita authorities and others when they asserted that the meat produced by the shochet was not kosher.[44] In not always successful attempts to regularize the application of religious legal requirements, the communities

39 Ibid., 179.

40 A. Wiener, *Die Jüdischen Speisegesetze* (Breslau, 1895); Steven Singer, "Jewish Religious Thought in Early Victorian London," *AJS Review* 10, no. 2 (1985): 195.

41 Albert M. Hyamson, *The Sephardim of England: A History of the Spanish and Portuguese Jewish Community, 1492–1951* (London: Methuen, 1951), 67; *A History of the Jews of England* (London: Jewish Historical Society/Chatto & Windus, 1908).

42 Hyamson, *The Sephardim of England*, 178–179.

43 Charles Duschinsky, "Jacob Kimchi and Shalom Buzaglo," *Jewish Historical Society of England–Transactions* 7 (1911).

44 "Jewish Law and Kosher Meat," *Times*, February 8, 1911; "Jews and Meat: The Rule in England," *Manchester Guardian*, July 12, 1907; "The Jews and Their Customs: Remarkable Libel Action," *Manchester Guardian*, February 26, 1904; "A Local Synagogue and Its Butcher," *Liverpool Mercury*, December 16, 1891; "Court of Exchequer, *Schott v Adler*," *Daily News*, December 15, 1868; "Monday came on in the Court of Common Pleas, before Lord Loughborough, and a special," *Times*, December 18, 1786.

of London (and United Kingdom) Jewry, sought to create an administrative structure that would deal with the difficulties arising from the evolution of an increasingly diverse population.[45] Boards of Shechita were established in each urban center in which there was a significant Jewish community, although the claims to monopoly jurisdiction by these Boards did not always sit well with all shochetim, or all rabbinical authorities.

Since its beginnings in 1760, the Board of Deputies of British Jews had provided the key institutional structure and support for the political struggles relating to the protection of Jewish rights and the battles for emancipation. Despite its history of intra-Jewish disputes, and periods of inaction in its earliest years, the Board of Deputies by the middle of the nineteenth century was the key voice for the Jews of London and the United Kingdom as the communities sought to concretize Jewish political, social, and religious equality, and to protect Jewish religious practice.[46] In 1804, the London Board for Shechita was formally established.[47] Local boards replicating its functions would later be established throughout England, particularly in larger centers such as Manchester, Leeds, and Liverpool.

The 1855 Prosecution of Shechita in London

It is perhaps not without irony that in the same year as the first prosecution for animal cruelty against the Jewish method in London, 1855, David Salomons, a leading figure in the fight for Jewish political emancipation, and the first English Jew to serve as a sheriff, magistrate, alderman, and Member of Parliament, became the first Jew to act as Lord Mayor of the City of London.[48]

In early October 1855, the RSPCA issued a summons against "Mr. Israel, the Jew butcher of Aldgate" for animal cruelty in the slaughter of an ox.[49] At the first appearance of the parties, "Mr. Israel" was represented by his solicitor's

45 Edelman, *The Jews in Britain, 1656 to 2000.*

46 Charles H.L. Emanuel, *A Century and a Half of Jewish History: Extracted from the Minute Books of the London Committee of Deputies of the British Jews* (London: Routledge, 1901).

47 Laws of the Board for Administering the Affairs of הטי״חש (London, 1880); Albert M. Hyamson, *The London Board for Shechita, 1804–1954* (London: London Board for Shechita, 1954).

48 Albert M. Hyamson, *David Salomons* (London: Methuen, 1939). Ten years later, another Jew (Sir) Benjamin Phillips, would become Lord Mayor. "Obituary," *Times*, October 10, 1889.

49 "The Police Courts: Mansion House," *Daily News*, October 6, 1855. The account is taken from this source.

clerk and the secretary, and prosecutor, of the Royal Society for the Prevention of Cruelty to Animals, Henry Rumsey Forster, acted for the prosecution. The charges, as would be the template in subsequent cases, related to casting, "by fastening the head of the animal to the ground by a ring through its nose, attached to a staple," and secondly to the killing of the animal itself, "having improperly cut its throat." Sir Peter Laurie acted as magistrate for the City of London.[50] He summoned the parties to a pre-trial meeting to learn more about the accusation because "the complaint had excited a great deal of curiosity, and given rise to much conversation in both Christian and Jewish communities." The Board of Shechita had entered into unsuccessful discussions with the RSPCA to explain the origin and nature of shechita and avoid, if possible, a prosecution.[51] In the earliest days of the case, the *Jewish Chronicle* had reported to the Jewish population that the Shechita Board itself had received the summons from the RSPCA.[52]

The representative of the RSPCA explained to Laurie their legal action attacked the Jewish mode of slaughtering. The ordinary mode of slaughter, through stunning, resulted in an immediate loss of all sense of pain, whereas the Jewish method caused the animal to endure ten to twelve minutes of suffering. This amounted to unlawful and cruel ill-treatment under the statute. Laurie interjected with his understanding that the Jewish method was grounded in religious law. While he made it clear that judicially he could form no judgment on the matter before the formal proceedings, he stated that he found the RSPCA's assertions that the Jewish practice was intentionally inhumane abhorrent in "present enlightened days." Forster replied that "his observations were made in no unfriendly spirit, and he felt convinced that the subject would be treated in the most liberal and impartial manner." The rhetorical field of judicial intervention in shechita was set and would not waver for the next fifty-eight years leading up to the prosecution of Rabbi Levitt in Halifax. The process of casting and of cutting the throat without prior stunning was cruel and inhumane according to the RSPCA, and there could be no distinction between the preparatory acts relating to casting the animal, and the cutting of the beast's throat. The Jewish mode of slaughter without stunning, unlike the

---

50 Laurie was himself a sheriff and later Lord Mayor of London. He was knighted in 1824. *London Gazette*, April 10, 1824, 580; Anita McConnell, "Laurie, Sir Peter," *Oxford Dictionary of National Biography*, http://www.oxforddnb.com.

51 Hyamson, *The London Board for Shechita*, 25–26.

52 "The Shechita Board," August 31, 1855.

ordinary mode, was in and of itself inhumane and cruel, but for the RSPCA no anti-Jewish sentiment informed this view.

The prosecution brought by the RSPCA was based only in concern for the animals and aimed at the elimination of cruelty and suffering. It did not in any way, shape, or form, target Jews, or Jewish religious practice, and was not motivated by anti-Judaic prejudice. Except that it most obviously did and was, since the goal of the summons was to punish the shochet for practicing the Jewish method of slaughter, from casting to slitting the ox's throat without prior stunning. Forster was aware of the fact, of which he informed the magistrate, that if the first cut did not kill the animal, it was forbidden under Jewish law to make another and permit the meat to be consumed by Jews. But no such failure or incompetence was at issue in this case. In the first documented instance of a cruelty prosecution against shechita in the Anglo-American world, there was no accusation by animal protection authorities that the killing was done in an incompetent fashion. This was a straightforward, unambiguous prosecution of shechita as cruel and inhumane. The case was set for a full hearing before Sir R.W. Carden, Lord Mayor, and Laurie ten days later.[53]

Both prosecution and defense were represented by barristers and "the justice-room was filled with persons interested in the question."[54] William Ballantine, one of the leading barristers of his day, acted for the defendants.[55] (Sir) William H. Bodkin, another leading lawyer, represented the prosecution.[56] In the formal proceedings, the accused shochet was named as Yankoff Cohen, rather than "Mr. Israel," but the charges remained unchanged. Thomas Reynolds, an English butcher, was charged with cruelty relating to the casting the ox and Cohen with ill-treatment "by improperly cutting its throat." Regardless of the prosecution's religious or ethnic biases or motives in bringing the case, the overt condemnation of shechita, if confirmed by the court, would

---

53 Sir Robert Walter Carden was a wealthy banker and stockbroker. He was active in politics and campaigned against vagrancy in his time as magistrate. Peter M. Claus, "Carden, Sir Robert Walter," *Oxford Dictionary of National Biography*, http://www.oxforddnb.com.

54 "The Police Courts: Mansion House," *Daily News*, October 17, 1855. The quotations from the case come from this account.

55 Thomas Seccombe and H.C.G. Mathew, "Ballantine, William," *Oxford Dictionary of National Biography*. http://www.oxforddnb.com. Ballantine was appointed serjeant-at-law and led many famous cases for both prosecution and defense.

56 James Mew and Nilanjana Banerji, "Bodkin, Sir William Henry," *Oxford Dictionary of National Biography*, http://www.oxforddnb.com. Bodkin served as a Member of Parliament and as judge. He was a leading advocate for poor law reform.

clearly have had a significant and possibly devastating impact on observant Jews in England.

Bodkin proudly proclaimed that the prosecution was not brought by anyone motivated by malignant motives or by pecuniary gain. The summons had been issued on behalf of "a body of gentlemen associated for the purpose of seeing that a class of God's creatures which were unable to complain themselves, were not made the subject of wanton cruelty, and that the very wholesome and humane provisions of the act, so far as their exertions would admit, were carried out."

The prosecution had not only the will of Parliament, wholesomeness, and humanity, but God on their side. The investigations into the theological origins of the practice under attack, however, had no place in a court of law as far as the RSPCA was concerned. Theology was good enough for the body of Christian gentlemen of the RSPCA, but was otherwise irrelevant to the case before the magistrates. What was before the court was the fact of which everyone was aware, "that the Jewish community had a peculiar custom with respect to the slaying of cattle for the subsistence of their people." That peculiar practice was before the bar of justice and it was the RSPCA's hope that the educated classes might turn their attention to these practices so that the prosecution, beyond the conviction of the accused, would serve the cause of humanity. The trial had both a penal and a pedagogical function in the mind of the RSPCA. They hoped that the broad educated classes would come to abhor the Jewish practice as they did, once the case had brought the inhumanity and cruelty of shechita to light. Perhaps they hoped that educated Jews would resile from the cruel and inhumane practices of their fellows, and reject the injunction of their holy obligations in the name of humanity.

While the prosecution placed some emphasis on the process of casting, and the delay in carrying out the killing of the ox, thereby causing it to remain in a painful position, the focus of their argument was on the distinction between "the modern and generally adopted system" of stunning, and the pain and agony inflicted on the beast under the Jewish system. Edward Dixon of the Metropolitan Police testified that he had attended, along with an RSPCA representative, on August 9, at the Aldgate slaughterhouses to witness three bullocks being killed by the Jewish method. He indicated that Reynolds had assisted in the casting and in holding the animal still before Cohen slit its throat. He reported that the animal's head was almost completely severed but that suffering for continued almost nine minutes. Upon careful cross-examination, he persisted in the view that poleaxing was quicker and caused less suffering than the Jewish method.

George Seaman, a farmer, butcher, and officer of the RSPCA, confirmed the policeman's evidence. Other butchers testified as to their experience according to which the stunning of animals resulted in a quicker, less painful, and more humane death than the Jewish mode. The witnesses Dr. Edward Lloyd of the Royal Dispensary and Dr. Pavey of Guy's Hospital supported this testimony scientifically by stating unequivocally that the Jewish method was not as humane as the stunning procedure. The evidence in this case, by eyewitnesses, experienced butchers, and medical experts set the stage for the narrative framing of all subsequent shechita prosecutions. Descriptions of animals thrashing after their throats had been cut, together with scientific opinion about the physiology of these movements, offered support for the RSPCA position that stunning before cutting resulted in but a short, almost instant, period of pain, from the original blow, whereas the Jewish method produced prolonged suffering and agony. The normal method was demonstrably more humane than the Jewish mode. RSPCA prosecutors in every subsequent anti-shechita case would maintain this position.

The defense opened and closed with an impassioned plea from Ballantine, who called no witnesses and satisfied himself with indicating to the court that, while he nominally represented the two accused, in reality he "represented the feelings and opinions of a very large body of the community" who viewed the prosecution "with a considerable amount of anxiety." The defense presented this as a case not of two defendants charged with individual acts of animal cruelty, but above all as a charge against the Jewish community and its religious practices. Ballantine touched on themes of Jewish piety and the humane commands of Jewish religious law. The ethical and theological idea of *rachamim* informed the barrister's presentation of the defense case. He placed emphasis on the command that cruelty be absent during the practice of shechita, under the norms of Jewish religious law, combined with the injunctions that animals be killed in a particular way. In making these assertions about the Jewish religion and the nature of the obligations in relation to shechita, Ballantine presented an overall position meant to underline how the case involved the whole of the Jewish community and its ability to adhere to the injunctions of their faith.[57]

57 "Letter to the Editor," *JC*, October 12, 1855.

This was followed by an attack on the RSPCA and its motivations in bringing such an action, the result of which would inevitably be to "raise a question of that kind, and to introduce, or take the chance of introducing, feelings of animosity which had been subsiding for years." While pulling up short of identifying antisemitic motivations behind the RSPCA prosecution, Ballantine's assertion referred to the anti-Jewish effects or potential effects of the legal case against the Jewish method. He made it clear that the consequences of the prosecution would go beyond the singular issue of shechita and could give rise to an upsurge in anti-Jewish feeling and actions. In the context of the still unfinished Jewish struggle for emancipation and equality, the barrister placed the RSPCA and its arguments against Jewish religious practice and belief on the regressive side of the debate. He attacked not the good faith of the RSPCA officials, but their good sense. "It was misfortune when great powers were entrusted to people not capable of exercising great discretion, and when the interests of large bodies were entrusted to the miserable management of those who did not understand how to deal with them."

Placing the fate of the Jewish people, or those among them who sought to obey the laws of kashrut, in the hands of the private organization, the RSPCA, led almost inevitably to this misuse of the criminal law powers under the statutory regime governing animal cruelty. The emerging public interest in ensuring freedom of religious practice and the reduction and elimination of social disharmony among the faiths, was, and would continue to be, ill-served by the privatization of cruelty investigations and prosecutions in the hands of those with a normative baseline of a limited vision of the interests of "that class of God's creatures which were unable to complain themselves." The miserable management of the RSPCA could never fully comprehend or act according to the fuller context of emancipation and freedom abroad at the time.

Jews were bound by holy law to follow the practice of shechita when choosing to consume meat and by that same law to abjure cruelty.

> It was known to those who laid the information that they considered themselves religiously and scrupulously bound, and that they considered they would be guilty of a great and heinous crime if they did not carry out strictly what was enacted in the book to which they looked up not only as their directory, but as complete and perfect revelation. The Jews could not conscientiously adopt any other mode than they were supposed to have adopted, and did adopt, in their slaughter-houses.

This is the earliest recorded legal plea before an English court justifying shechita in theological terms in an animal cruelty prosecution. In legal history, it stands as the clearest exposition of the Jewish legal and religious position. The Jewish method of slaughter was commanded by holy obligation, and to act as the RSPCA demanded by stunning according to the so-called normal method, would compel those of the Jewish faith to do something against their most fundamental beliefs. The RSPCA legal action, no matter its motivation, was an attempt to criminalize a core belief and practice of London's observant Jewish community.

Ballantine argued that it could surely not have been Parliament's intent, in an era of increasing tolerance and given the significance of social and religious harmony, in passing animal cruelty legislation, to criminalize Jewish religious practice. "They did not interfere with their neighbours, and all they asked was to be let alone." He concluded by urging the magistrates to "pronounce a judgment which would have the effect of putting a stop to such prosecutions for the future." The plea was partially reflective of social, political, and legal reality, since the formal emancipation of the Jews of the United Kingdom was not yet fully in place. It was consistent with important aspects of liberalism within the spirit of the times, both in its assertion of tolerance in the public sphere and in relation to private religious practice. In its appeal to the binding nature of holy scripture in the time of mid-nineteenth-century Christian revival, the idea that one's actions should be based on the will of God would have struck a resonant chord with many from different parts of the Protestant community.

After a ten-minute recess, Sir Peter Laurie delivered the decision. The court's duty was to interpret the relevant act of Parliament and determine if the prosecution had established deliberate acts of cruelty. While the general principle must remain that religion alone could not be a defense to cruelty, on the facts of the case no offense had been proven. There was no indication that the Jewish method demanded the infliction of deliberate and wanton pain "without a lawful and justifiable object." The Jewish method of killing oxen was no different than that used by "every Christian butcher in killing calves, sheep, and pigs." It would be inappropriate for the court to adopt the RSPCA's narrow ideal of stunning in all slaughtering of animals for food, especially when non-stunning was widespread in the industry itself. To accept the RSPCA opposition would mean that no butcher could carry on his trade. "I should be interfering with the ordinary and usual business of life and, above all, meddling with matters of faith and religious opinion on the subject of ceremonial observances, which we are all bound to respect so long as they do not involve any breach of the common law."

"The decision was followed by cheering." In this first instance of a prosecution of those engaged in the practice of shechita, the Jewish community of London enjoyed an apparently unequivocal victory. That victory was obtained as the result of a legal argument grounded in assertions of freedom of religious observance from the defense, but those assertions were only partially echoed in the magistrate's reasoning. Future cases would involve ideas of scientific progress, and the RSPCA developed a somewhat more sophisticated prosecution strategy to which Jews would be forced to respond. The Jewish community would be compelled to fight science with science, the physiology of pain with the physiology of pain. This was not the case in the London prosecution of 1855. The Jews of London fought back with scripture, religious obligation, and a spirit of emancipation. The court of the City of London heard a legal plea about Jewish religious freedom and RSPCA intolerance and ignorance that set out the terms of the debate over shechita that would be unfortunately obscured and often forgotten in subsequent proceedings. The victory was not unequivocal. Laurie placed significant weight on the ordinary practice of animal slaughter without stunning that was then current throughout the land. As that practice evolved with killing and stunning technological developments, the discourse would move into different and more problematic terrain.

While later prosecutions would serve, sometimes intentionally, sometimes unintentionally, to obscure the religious aspects of these cases, there is no doubt that in 1855 the Jewish communities and others saw that fundamentally conflicting normativities were at play. The case received significant attention in the press, both in London and elsewhere.[58] In the minds of non-Jewish observers, the case involved a clash between the RSPCA and the Jewish communities and it was described and understood as an outright attack on the Jewish method of slaughter. It was "a very unwise departure from their (RSPCA) proper object, and a most unwarrantable interference with usages founded upon religious conviction."[59]

58 "The Jews and the Society for the Prevention Cruelty to Animals," *Liverpool Mercury*, October 13, 1855; "Alleged Cruelty to Animals–The Jews," *Bristol Mercury*, October 20, 1855; "The Jews and the Society for the Prevention of Cruelty to Animals," *Sheffield and Rotherham Independent*, October 20, 1855; "Jewish Mode of Slaughtering Cattle," *Observer*, October 21, 1855.

59 *The Patriot*, quoted in "Legality of Jewish Customs," *JC*, October 26, 1855; "The Christian and Jewish Modes of Slaughtering Animals," *Daily Telegraph*, October 24, 1855; "The Society for the Prevention of Cruelty to Animals and the Jews," October 10, 1855.

The 1855 case could be, and was, experienced by the Jewish community as a great victory for religious freedom. Many years passed before another case would be brought against the Jewish method, but that did not mean that shechita was immune from attack. The Board of Shechita kept a watching brief on the matter throughout the 1870s and the 1880s, when prosecutions took place outside London.[60] What could not be gained by the pursuit of criminal cases could be sought by animal protection advocates in the form of legislative change and local regulation of slaughtering practices.[61] The RSPCA and its allies were not quiescent. At the same time, and in parallel, internal Jewish debates over shechita, the licensing authority over shochetim, and religious authority generally, continued in the years following Laurie's decision.[62]

60 Hyamson, *The London Board*, 36, 40.

61 Report of the Committee Appointed by the Admiralty to Consider the Humane Slaughtering of Animals; Raphael Langham, *250 Years of Convention and Contention: A History of the Board of Deputies of British Jews, 1760–2010* (London: Vallentine Mitchell, 2010), 101–103, 156–159.

62 Benjamin J. Elton, *Britain's Chief Rabbis and the Religious Character of Anglo-Jewry, 1880–1970* (Manchester: Manchester University Press, 2000); Eugene C. Black, *The Social Politics of Anglo-Jewry, 1880–1920* (Oxford: Basil Blackwell, 1988).

CHAPTER 4

Manchester Shechita: Criminalizing Orthodox Judaism 1878 and Beyond

Manchester Jewry and Shechita

Twenty-three years after the Mansion House case, the RSPCA in Manchester brought a prosecution for animal cruelty against Messrs. William Seymour, William Shaw, and William Morris, as well as Marks Morris. Seymour, Shaw, and William Morris were "English journeymen butchers." Marks Morris is described in the newspaper accounts as "second reader at the Old Synagogue" and as holding "some semi-ecclesiastical office in connection with the Jewish synagogue."[1] The 1871 census records his profession as "second reader in the Jews synagogue."[2] The four men were charged with "cruelly beating, ill-treating, abusing, and torturing a bullock." The alleged cruelty related to the preparation of the animal for slaughter, in other words the process of casting whereby the animal was fastened to rings attached to the wall for some time, before its throat was cut. On the surface, this case was concerned with the way in which the animal was placed in a position to be killed by the Jewish method and the main actions were those of the English butchers who had assisted the shochet.

By the time of the prosecution, there had been an active Jewish community in Manchester for over one hundred years. Following the establishment of a group of Jewish merchants from Liverpool in Manchester, the community had experienced years of growth and expansion, as the city became a center of the industrial revolution. In the ten-year period from 1861 to 1871, a few years

1 "The Jewish Mode of Slaughtering Cattle," *Jewish Chronicle*, July 19, 1878.

2 Census Returns of England and Wales, 1871, RG10, Piece 4062, Folio 93, Page 34, GU roll 846345, National Archives.

before the RSPCA sought to criminalize shechita again, the number of Jews had increased from 1,755 to 3,444.[3] The next census, enumerating the population three years after the prosecution, shows the dynamic nature of immigrant expansion in the industrial heartland of England's north, as the arrival of eastern European migrants had swollen the city's Jewish population to some 7,000.[4] The Manchester case of 1878 is important not just as a concrete embodiment of the attempts in the Victorian and early Edwardian eras by the RSPCA to use the existing animal cruelty legislation to mount a series of attacks on the Jewish method of slaughter, but because it allows for a move away from the center of Anglo-Jewry in London, to the circumstances and experiences of Jews in the provinces.[5] The idea of a distinct local experience for Jews outside London has only recently moved from the shadows of the capital. "It is neither original nor controversial to state that (except for the medieval period) the historiography of the Jews in England has hitherto shewn little interest in what went on outside the capital."[6]

Since Roth's early study, greater attention has been paid to the specificities of what Kushner has characterized as "place, locality, and memory" in English, and British, Jewish history.[7] The RSPCA's attack on shechita extended to Scotland. The lawyer for the Jewish community in the Aberdeen case in 1893 invoked the acquittal of Yankoff Cohen in London in 1855 as part of collective legal memory. But no mention was made of the experiences of provincial English Jewry with the anti-shechita movement, including the 1878 Manchester case.

By the time of the prosecution of the English butchers and the shochet, the Jewish community was significant and thriving. From the earliest days, the community developed a complex, effective system of collective institutions for self-governance, the provision of welfare, and educational services for the Jews

3 Bill Williams, *The Making of Manchester Jewry, 1740–1875* (Manchester: Manchester University Press, 1985), 355; Lipman, "A Survey of Anglo-Jewry," 183–184.

4 Laura Vaughan and Alan Penn, "Jewish Immigrant Settlement Patterns in Manchester and Leeds 1881," *Urban Studies* 43, no. 3 (2006): 658.

5 Aubrey Newman, "Provincial Jewry in Victorian Britain: A Report," *Jewish Historical Society of England–Transactions* 25 (1973–1975).

6 *The Rise of Provincial Jewry: The Early History of the Jewish Communities in the English Countryside, 1740–1840* (London: Jewish Monthly, 1950), 11.

7 Tony Kushner, *Anglo-Jewry Since 1066: Place, Locality, and Memory* (Manchester: Manchester University Press, 2009); Aubrey Newman, ed. *Provincial Jewry in Victorian England* (London: Jewish Historical Society of England, 1975).

of Manchester.[8] It was large and diverse enough that the community had suffered disagreements, splits, and schisms, and by 1857 there was a large Reform congregation competing with traditional Orthodoxy.[9] Differences of opinion, belief, and practice existed within the Orthodox community. Important aspects of the history of Manchester Jewry were characterized by ongoing struggles with the center, and particularly with the office of the Chief Rabbi in London, and local insistence on various aspects of religious autonomy.[10] At the heart of many of these controversies was the issue of shechita—the right to name and approve shochetim, to certify products as kosher, and, an important matter for smaller regional communities, the right to levy a tax on kosher goods. The provision of kosher food had long been central to the Jewish life of Manchester and had been located primarily in the synagogue.

> The synagogue was the local custodian of Jewish observance, providing services and facilities which made it possible for Manchester's Jewish residents to live and die within the Mosaic dispensation. The congregational *shochet* killed for the community according to the traditional rules of *Kashruth,* and supervised the sale of kosher meat at the shops of licensed butchers, who paid a tax to the synagogue for his services; when no suitable abattoir was available, the congregation rented a slaughtering yard of its own.[11]

Manchester's Jews would pass through all the difficulties relating to complex shifts and splits within the different congregations as more Jews with different versions of observance arrived in the industrial capital from central and eastern Europe. As in London, religious authority and monopoly over kashrut and shechita would give rise to struggles, culminating in Manchester's own set of kosher libel cases.[12] The Manchester Board of Shechita, the institutional manifestation and embodiment of an attempt to unify and clarify the governance of

8 Rainer Liedtke, *Jewish Welfare in Hamburg and Manchester c. 1850–1914* (Oxford: Clarendon Press, 1998); *They Came from the Haim: A History of Manchester Jewry from 1867* (Manchester: Jewish Social Services Greater Manchester, 1995).

9 P. Selvin Goldberg, *The Manchester Congregation of British Jews, 1857–1957* (Manchester: Congregation of British Jews, 1957).

10 Elton, *Britain's Chief Rabbis*; Geoffrey Alderman, "The British Chief Rabbinate: A Most Peculiar Practice," *European Judaism* 23, no. 2 (1990).

11 Williams, *The Making of Manchester Jewry,* 32–33.

12 "Jewish Customs," *Manchester Guardian,* December 16, 1868.

Magistrate, Francis John Headlam, who served for over thirty years as a magistrate in Manchester. "His long occupancy of that position had since made him one of the most familiar figures in Manchester and his name one of the best known."[18] A prominent member of the Manchester legal establishment (Sir) William Cobbett, represented the prosecution. Cobbett twice served as head of the Manchester Incorporated Law Association and was particularly well known for his community service and active role in the building of the new facilities of the Manchester Royal Infirmary.[19] The presiding judge of the Manchester Assizes described Cobbett, upon his death, as "an ornament to the legal profession." The local County Court judge mourned the loss of "a great lawyer and advocate, a great citizen and man of affairs, and a great gentleman."[20]

J. H. P. Leresche appeared for the defendants. Another leading member of the Bar, he later served as Deputy Recorder of Manchester and as a stipendiary magistrate. He was "esteemed and respected by all his brother magistrates," and by the police.[21] Clearly, this was no minor legal matter.

Cobbett commenced his case by demonstrating the "practice commonly adopted in this country" for slaughtering cattle.[22] The scene was set from the beginning—there was a common method involving stunning before killing "there being continuous insensibility throughout the operation." By way of contrast, there was the Jewish method of killing without stunning, during which there was sensibility, suffering, torture, and cruel ill-treatment of the beast. Cobbett and the RSPCA appeared to have departed from the traditional attack on shechita with its core focus on the killing of the animal without prior stunning. The lawyer explained, the focus in this case, and the reason the three English butchers stood accused, was on the process leading up to the killing. The prosecution alleged and would prove cruelty not in the slitting of the animal's throat, but in the preparation of the animal, placing it in the position

18 "Mr. F. J. Headlam," *Manchester Guardian*, March 17, 1908; "Mr. Headlam: References by the Justices," March 18, 1908.

19 "Sir William Cobbett," *Manchester Guardian*, November 27, 1926.

20 "The Late Sir William Cobbett: Notable Legal Tributes," *Manchester Guardian*, November 30, 1926; "Sir William Cobbett: Funeral at Wilmslow," December 1, 1926.

21 "Memorial Service: Mr. J. H. P. Leresche," *Manchester Guardian*, March 7, 1894; "The Late Mr. Leresche," March 8, 1894.

22 "The Jewish Mode of Slaughtering Animals: Singular Charge of Cruelty," *Manchester Guardian*, July 12, 1878. All quotations are taken from this account. Magistrate and police court records contain no trace of the case. Tony Lees, Archives Officer, Manchester Central Library, personal communication, March 22, 2017.

with its neck properly exposed to the *hallaf*. The allegation was that casting as practiced in Manchester was cruel and illegal.

Cobbett asserted that the objection was not to bleeding the animal to death, since it was "not of deaths a very painful death." The difference between bleeding following stunning and bleeding without stunning was "not very great." The illegal cruelty occurred in preparing the animal to be slaughtered. With its focus on casting, it would appear that this was not an attack on shechita. Nothing in Jewish law compels a particular method of casting. The general criterion that the animal not be injured so as to render it unfit to be properly killed and later consumed according to Jewish law, and the practical requirement that it be placed in a position permitting the shochet to kill it with the single quick cut, apply, but there were no specific religious commands about the process of casting itself.

Cobbett provided the court with a detailed description of the way in which casting was carried out in Manchester by the three accused, including graphic detail concerning the immobilization of the cow, the twisting of the animal's neck into position by a process of stretching manually and with the assistance of ropes and poles. According to the RSPCA, this resulted in pain being inflicted on the animal throughout, a process of near suffocation, with scientific proof of extreme suffering. The barrister introduced photographs, not of the particular instance before the court, but of casting using the same method, to highlight and give a physical sense of what was involved to the magistrate. The relatively new technology of photography was used by the RSPCA to illustrate its case and to indicate that its ideas of cruelty and ill treatment were grounded not in religious prejudice but in scientific fact.

The argument was buttressed by testimony—by eyewitness evidence from an RSPCA inspector, a police constable, and a veterinary surgeon, each of whom had been present when the animal had been killed, and each of whom offered evidence and accounts of what had happened to support Cobbett's opening statement. Five further veterinary surgeons testified that animals suffered unnecessary pain when placed in the position followed in this instance and casting of this kind was used. The prosecutor summed up the case against the accused in terms that carefully circumscribed the RSPCA position and set out the position that this was not an attack on shechita. "The Society which he represented had no desire to interfere with that method of slaughter; all they wished was, that before it was bled to death the animal should not be put to unnecessary pain."

The prosecution was one that targeted a particular mode of casting the animal, and not one that sought to eliminate killing without stunning.

The argument in this case was that the RSPCA admitted that the suffering caused in the Jewish method of slaughter was perhaps only minimally greater than the suffering involved in bleeding following stunning, and did not rise to the level of cruelty envisaged within the statutory prohibition. In this first recorded prosecution following the 1855 London case, the RSPCA had abandoned its insistence that killing without stunning was in and of itself a clear infliction of illegal suffering on the animal. It was now focused only on the casting aspect of slaughter. Cobbett emphasized that the RSPCA did not imagine that the process of casting in another way that did not result in such suffering would violate any of the beliefs of the Jewish community.

What passed unnoticed is the RSPCA inspector's testimony that they attended the slaughterhouse "in accordance with instructions." This was not a random or an ordinary inspection of a slaughterhouse. A police constable and a veterinary surgeon accompanied the inspector under specific instructions to witness a killing according to the Jewish method. While Cobbett could claim that this was not an attack on the mode of killing without stunning, the evidence from his own witnesses indicates that the Manchester RSPCA targeted shechita for investigation and prosecution. If they limited themselves to the casting part of the slaughtering process, this was perhaps due more to the fact that they were compelled to admit that the act of killing with a swift single blow from a perfect *hallaf* could not amount to cruelty under the statute, than to any acceptance of slaughter without prior stunning.

Leresche mounted a defense based in part on Jewish religious belief and in part on the wording of the statute and the requirements of the law for a criminal conviction. He repeated the argument that had been advanced by Cobbett, that the statute, as interpreted, required a specific form of intention. As the prosecutor had explained,

> ... it had been decided over and over again that where a particular act of torture had been inflicted in order to make the beast or bird or whatever it was fit for the purpose of mankind, that that act was lawful, but that there was a distinction to be drawn between those cases and cases where cruelty had been done for purposes not necessary for the good of mankind, but for the advantage, pleasure, or profit of particular individuals.

Leresche argued that the statute permitted any act that was not intentionally cruel. The animal had been cast and killed with the intention of providing meat that would comply with Jewish religious belief and law. Such a case could never be considered to have been carried out with the wanton cruelty required under the law. Leresche moved on to an argument that echoed Ballantine's advocacy in London in 1855. Morris was the shochet of the Jewish community, approved by the Chief Rabbi as possessing the necessary skills and knowledge to carry out the Jewish mode of slaughter. That method of killing was one of the "obligations of divine law" that the Jewish community was bound to observe. Belief in mercy and kindness to all creatures was a core element of Jewish existence, and the method of killing was part of that ethos. Leresche presented his overarching case in the context of Jewish religious belief and obligation, as well as in an assertion that Jewish ethics prohibited the same acts of wanton cruelty outlawed by the statute. The central value of *rachamim*, compassion and mercy, has always had an overarching presence in responses to allegations of Jewish cruelty in SPCA prosecutions and in all Jewish texts promoting an ethical regard for animals.[23] For the defense, this case was not about casting; it was case about Jewish religious duty. "It was remarkable that they should now be charged, not only with breaking the law of England, but with violating the Mosaic law under which they had lived for centuries."

The barrister made his legal argument about the position of the Jewish community and its rights and obligations. This was not a casting case; it was an attack by the RSPCA on shechita. Leresche made no mention of the English butchers in his plea. He referred only to Marks Morris and his qualifications and duties as shochet to ensure that the animal was not damaged or injured prior to slaughter and was killed according to Jewish law. If this were not a case against shechita in the guise of a casting prosecution, one might ask why Morris figured among the defendants. He played no part, according to the multiple eyewitnesses who testified for the RSPCA, in casting the animal, the series of events described so graphically by Cobbett. Morris had entered the room and simply slit the throat of the cow, following the method prescribed by Jewish law. Unless the accusation was one involving Morris having ordered and supervised the casting, as an accomplice, his only act was one with which the prosecution had specifically stated it had no legal objection.[24] Allegations about the cruelty of a particular casting method disguised a more foundational attack against shechita, notwithstanding the RSPCA barrister's claims to the contrary. The Manchester prosecution included Marks

23 Aaron S. Gross, *The Question of the Animal and Religion: Theoretical Stakes, Practical Implications* (New York: Columbia University Press, 2015).

24 The statute imposed accessorial liability only in cases where the accused "caused" or "procured" the ill-treatment of the animal. Cruelty to Animals Act, 1849, § 1.

Morris because it was assumed that what was under attack was a Jewish process in which the Jew would have been in command. In reality, throughout the period of anti-shechita prosecutions, the lack of Jewish control over casting had been a major concern of religious officials.

Leresche called a series of witnesses, including experienced butchers and slaughtermen who testified that casting was the best way to place the animal in a position in which it could be properly killed by the shochet. Dr. Arthur Gamgee, professor of physiology at Owens College (later the University of Manchester) testified that he had witnessed the Jewish method and that it was a humane way of killing that inflicted a minimal amount of pain on the animal. Casting was necessary to allow the single cut. Casting as described was a familiar process, used in the performance of surgical operations on large animals by veterinarians.

The defense case followed what would become a dual strategy, the first part of which was an attack on the science of pain, in which experts testified that the Jewish method was at least as painless, or only as painful as, the normal method of killing by stunning. Despite the earlier insistence by the prosecution that this was not an issue, it did feature both in the RSPCA evidence and that brought by the defendants. Despite protestations to the contrary, the RSPCA case did attack shechita as a whole. Because casting is essential in the Jewish method, so that the animal's neck can be properly presented to the shochet's blade, it is difficult, if not impossible, to separate the casting of the animal from the Jewish method. The inclusion of Marks Morris as an accessory can only be explained in this instance as the necessary consequence of the prosecution understanding the process of slaughter under the laws of shechita as a single event from casting to killing, and as one under Jewish control.

No method of casting is religiously obligatory and the Jewish community was constantly concerned with ensuring the humane nature of this part of the process. But the real underlying objection in all the cases brought by the RSPCA involves a rejection of killing without stunning. Casting is unnecessary in the normal method because the animal is unconscious or dead and it can be cut and bled without difficulty. Casting is required in the Jewish method precisely because the animal is not stunned. What this means is the RSPCA can (and did) only prosecute because the Jewish method of slaughter does not involve stunning. This was an indirect attack, as the Jewish community and its barrister, understood, on shechita itself.

The Police Court judge quickly reached his decision: that "there was no intentional cruelty in the minds of the defendants. The custom was one which had lasted a long time. As the evidence was contradictory he did not think there should be a conviction."

The defendants benefitted from the burden of proof in criminal cases, and from their positive demonstration on the evidence that they lacked the cruel intentions required for the offense. The case was treated as one that fell in the end within the normal course of events in the life of the Manchester court. There was a reasonable doubt on the issue of suffering raised by competing expert testimony, and there was little, if any, evidence of the wanton ill-treatment of the animal demanded under the statute. For the legal history of cruelty prosecutions in relation to shechita, the defense case was presented under a clearly articulated understanding that the RSPCA charges, despite the protestations of Cobbett, targeted Jewish religious belief and practice in an unambiguous way. Leresche argued that Morris was a duly certified shochet acting under the ambit of his religious duties as a representative of the Jewish congregation. He was designated to provide meat that could be consumed by the congregation in accordance with religious law. The defense barrister asserted that Jewish law and ethical teaching insisted on treating all creatures with dignity and humanity. The summons was dismissed in large part because Leresche adopted an unequivocal understanding of what was at stake before the Manchester Police Court and followed as far as possible the legal arguments advanced by Ballantine years earlier in the London case.

The headlines in the local press coverage, as well as in the *Jewish Chronicle*, indicated that there was a broad understanding that the case, as Leresche's plea set out, was not an ordinary one of cruelty, or one aimed only at the preparatory stages of casting. Jews and non-Jews understood that the RSPCA was aiming at the Jewish method in its prosecution.[25] In subsequent years in the north of England, a steady campaign continued, aimed at informing the public of the cruelty of the Jewish method, and highlighting the constant scientific improvement of humane stunning and slaughtering techniques, to be contrasted with the far less scientific inhumane Jewish mode.[26] The Jewish community fought back with their own continuous assertions of the essential humanity of shechita, supported by the latest scientific study.[27] Acquittals in cases brought against

25 "The Jewish Mode of Slaughtering Animals"; "The Jewish Mode of Slaughtering Cattle," *Manchester Times*, July 13, 1878; "The Jewish Mode of Slaughtering Cattle," *JC*, July 19, 1878.

26 "The Slaughtering of Animals," *Manchester Guardian*, September 6, 1893; "Alleged Cruelty in the Slaughterhouses: A Leeds Butcher's Views," *Leeds Mercury*, April 4, 1896.

27 "Jewish Butchering," *Leeds Mercury* February 29, 1896.

shochetim, scientific argument in support, emancipation and liberalized attitudes, thousands of years of religious practice and divine command, none of this could stop the RSPCA and its allies in their campaigns against the Jewish method of slaughter. The Manchester shechita case was not the first attempt by the RSPCA to bring criminal sanctions to bear upon the practice of the Jewish method of slaughter, nor would it be the last.

CHAPTER 5

The Birkenhead Shechita Cases

Prosecuting the Jews of Liverpool

The next prosecution for cruelty in relation to Jewish slaughter took place in Manchester's Lancashire rival city, Liverpool, seventeen years after the case against Marks Morris was dismissed and forty years after the London summons of Yankoff Cohen had failed. In early October 1895, the RSPCA again charged a "Jewish butcher," Joseph Myers, with animal cruelty. The case occurred in Birkenhead, across the Mersey from Liverpool. The prosecution alleged that Myers had left a bullock in a painful position during the casting phase of the slaughter, had only severed two of the four principal blood vessels when slitting the beast's throat, and had intervened to prevent the English butcher from dispatching the bullock by means of a poleax.[1] The RSPCA Inspector William Pocock charged Myers with "causing an ox to be cruelly illtreated."[2]

While the Jewish method of slaughter was specifically identified by the RSPCA and their barrister, the case focused, on the surface, on the poor execution of the killing itself. The two parts of the slaughter process singled out were the casting, as in the Manchester case, and the slitting of the animal's throat, which, according to the RSPCA, had been carelessly done by Myers. The charge was not one that sought to criminalize the practice of shechita itself, but instead it invoked ideas of negligence, carelessness, and incompetence in carrying out the killing according to the Jewish method, and the cruelty and ill-treatment of the animal that were the necessary consequences of this incompetence.

The RSPCA on Merseyside had a long, active history by the time of the Myers prosecution. Moss records the creation of the Society for the Suppression

1 "The Slaughter of 'Kosher' Meat: A Question of Religion and Cruelty," *Weekly Standard and Express* (Blackburn), October 5, 1895.

2 Birkenhead Justices: Police Information and Complaints Books, B/13x/142/20, Wirral Archives Service, Birkenhead.

and Prevention of Wanton Cruelty to Animals in the city in 1809.[3] While that organization struggled and fell into inactivity, by the 1840s, Liverpool became one of the earliest auxiliary societies of the London SPCA. Despite occasional internal disputes, the organization had a long history of activity in animal welfare.[4] The work of the Wirral branch in supervising animal welfare at the Birkenhead lairages was rewarded with a recurring grant of twenty-five pounds from the Mersey Docks and Harbour Board.[5]

The Jewish community of Liverpool had an even longer existence in the northwest of England. In his history of provincial Jewry, Roth states: "The community of Liverpool is the only important one in the whole of England, outside London, the history of which has been comprehensively written."[6] According to Roth's brief account, traces of an established Jewish community could be found in Liverpool in the 1750s, and firmer evidence of the community's formalized rules of governance, together with the important acquisition of a Jewish cemetery, date the Jewish presence to the last days of the eighteenth century.[7] While the community would not keep up with the dynamic growth of its Lancashire neighbor, as late as 1851, the Jewish population of the city was still greater than Manchester, and Liverpool was the largest Jewish settlement outside London.[8] While the population would ebb, flow, and ultimately decline, there has been a permanent Jewish presence for over 250 years.[9]

Liverpool's position as a port meant that the city continued to exist as a major center for trade, and the city's Jewish community was able to take advantage of the economic situation. As David Cesarani pointed out, however, the city flourished as a port for a significant part of its history thanks in no small measure to the Atlantic slave trade. This did not lead Cesarani to accuse Jews of intimate involvement in that trade. He pointed out that the vibrancy of the Liverpool economy as a whole cannot be understood outside the context of the slave trade. Jews, like many others, were implicated in, and profited from, the North Atlantic/Caribbean nexus indirectly. "For, although Jews arrived too

3 *Valiant Crusade*, 20.
4 "Liverpool (Branch)," *Animal World*, December 1875, 179; "Liverpool Branch of R.S.P.C.A.," March 1873, 35.
5 "Birkenhead," *Animal World*, May 1895, 80.
6 *The Rise of Provincial Jewry*, 82.
7 Bertram Benas, "A Survey of the Jewish Institutional History of Liverpool and District," *Jewish Historical Society of England–Transactions* 17 (1951–1952).
8 Lipman, "A Survey of Anglo-Jewry, 1851," 185.
9 Mervyn Goodman, "The Jewish Community of Liverpool," *Jewish Journal of Sociology* 38, no. 2 (1996).

late and had nowhere near enough capital to invest directly in slave ships before the slave trade was abolished in 1807, they were enmeshed in the local economy which was indubitably driven by the 'triangular trade'. Indeed, anything else would have been impossible."[10]

After the abolition of the slave trade, Liverpool continued to flourish as a major British seaport. Of key importance in relation to the prosecution of shechita in late 1890s, Liverpool engaged in trade across the Atlantic and the Irish Sea. Ireland was a major source of cattle (as well as sheep, pigs, and horses), and the city became a major center in the cattle trade, and a large animal holding and slaughtering facility, the lairages at Birkenhead, was central to the commerce in meat and animals. The transport of cattle by sea caused concern and outrage among those dealing with animal welfare.[11] The Liverpool RSPCA focused on the conditions in which animals were kept during their transport across the Irish Sea and in 1895 formed a special committee together with the Manchester RSPCA, with the involvement of other interested parties and beneficiaries of the Irish animal trade, the Dublin and Cork SPCAs from Ireland, and the Glasgow RSPCA from Scotland, and the Glasgow Fleshers' Association, Dublin Cattle Traders' Association, as well as the Liverpool Leeds and Bradford Butchers' and Meat Traders' Association.[12] Birkenhead had been for some time a point of focus for the RSPCA.

The Jewish mode of slaughter was practiced at Birkenhead both for the local Jewish community, and to serve the larger London market. At the time of the Birkenhead prosecution of Joseph Myers, the control of shechita at the lairages remained in the hands of local synagogal authorities. The centralized governance of the Jewish method of slaughter in the Liverpool area was still two years away.[13]

> The Liverpool Shechita Board was founded in 1897. Hitherto *shochetim* had been congregationally appointed, but as the Jewish population increased the situation became unsatisfactory in view of a number of

10 "The Jews of Bristol and Liverpool, 1750–1850: Port Jewish Communities in the Shadow of Slavery," in *Jews and Port Cities 1590–1990: Commerce, Community and Cosmopolitanism*, eds. David Cesarani and Gemma Romain (London: Vallentine Mitchell, 2006), 142.

11 I.M. Greg and S.H. Towers, *Cattle ships, and our meat supply* (London, 1894).

12 RSPCA Liverpool Branch, Committee Minute Book 1895–1903, October 8, 30, 1895, 179 ANI 1/3; Minute Book of Sub Committee re Irish Cattle Traffic 1895, 179 ANI 5/1, Liverpool Record Office (LRO).

13 Kashrut Commission September October 1895–96, OHC/21/105; Shechita Board Minute Book, 296 KAS 1/7, LRO.

> unauthorised functionaries. The position became even more difficult by the development of local *shechita* for the supply of meat for London. The Shechita Board thus became one of the most important organisations of its kind in the country.[14]

Shechita remained on the local and international agendas of animal protection organizations.[15] Local control by the different synagogues entered into a direct confrontation with the local RSPCA, before local magistrates, in a context of national and international legal significance.

The Prosecution of Joseph Myers

Joseph Myers stood trial before the Birkenhead Police Court, presided over by Joseph Craven, J.P., and G. R. Clover. Craven was a long-serving magistrate in the Police Court, and was a leader of the local community, serving on the board of the local industrial school and having been involved in the incorporation of the City of Birkenhead some years earlier.[16] Glover likewise was a longtime member of the local court, although he was perhaps better known as the head of a prominent shipbuilding firm.[17] The case was prosecuted for the RSPCA by barrister R.F. Colam, who appeared throughout the country for many years as the representative of the animal protection society.[18]

Other criminal cases against shochetim both before and after the 1895 Birkenhead case had excited the interest of the local Jewish authorities and the broader national and even international Jewish communities. Myers did not appear in Court until the prosecution case was well under way, and he was unrepresented.[19] It could appear that this was a case that concretized the disorganized and decentralized nature of the practice of shechita in Liverpool in the days before the Shechita Board was created. Later, the defendant's version of

---

14 "A Survey of the Jewish Institutional History of Liverpool," 32; N. Kokosalakis, *Ethnic Identity and Religion: Transition and Change in the Liverpool Jewry* (Washington, DC: University Press of America, 1982), 136.

15 "Latest Russo-Jewish News: Jewish Mode of Killing Cattle," *JC*, May 31, 1895.

16 "Death of Ald. Willmer, Birkenhead: An Interesting Career," *Liverpool Mercury*, October 16, 1897; "The Late Mr. Joseph Craven," April 6, 1897; "Deaths, Craven," April 3, 1897.

17 "Funeral of Mr. G. R. Glover," *Liverpool Mercury*, November 3, 1900.

18 "Obituary: Mr. R.F. Colam, K.C.," *Times*, May 5, 1942.

19 "The Jewish Method of Slaughtering Cattle," *Liverpool Mercury*, October 4, 1895. All quotations are taken from this account.

the story came to light in the pages of the *Jewish Chronicle*, first in the form of a letter from one Isaac Tumpowsky of Wolverhampton.[20]

Tumpowsky had been visiting Liverpool and took the time to interview Myers, whom he describes as the "Rev. Mr. Myers (Shochet)." The convicted butcher explained that the trial took place during Sukkot, and that he had earlier requested an adjournment because of the religious holiday. The Jewish community was not present, and Myers had no defense, because the trial was in essence sprung on him at a time when Jews were observing a lengthy holiday period.

The absence of Jewish community representation is even more regrettable given the opening speech by Colam, who made it clear to the court that the case brought by the RSPCA directly targeted the practice of shechita. "Mr. Colam said that this was an important case, because it presented for the first time before the court the consideration of the method of slaughtering animals adopted by persons of the Jewish persuasion."

Unless the RSPCA barrister was referring only to the Birkenhead Police Court, he was displaying a surprising level of disingenuousness or a lack of institutional memory. The RSPCA had brought similar prosecutions in London in 1855 and in Manchester in 1878. Two years previously, the case against Alexander Zamek and Rabbi Littman had been heard in Aberdeen. There was a long history of RSPCA agitation against shechita and a history of legal actions. That this case was part of that ongoing campaign became more evident as Colam voiced the general respect that the RSPCA had for all religious "prejudices," before turning to an outright condemnation of Jewish cruelty. Even "under the best and most favourable circumstances the method adopted was necessarily a very cruel one." The RSPCA barrister recognized the limits imposed on the prosecution by the religious obligations informing the Jewish method and the legal requirements in relation to the wanton cruelty necessary under the statute as interpreted and applied in English courts. A general and overarching legal condemnation of shechita was not permitted. As in Manchester, the RSPCA was targeting not the generalized practice of the Jewish method, despite Colam's introductory remarks to the contrary. The argument was that Myers had been at fault in failing to carry out the killing of the ox according to the strictures of Jewish practice.

There can be little doubt that the principle enunciated is legally sound. If Myers killed an animal outside the rules of shechita, and in the process caused

20 "The Birkenhead Decision Upon Shechita," October 18, 1895.

undue suffering and cruelty to be inflicted, the prosecution could proceed as a normal cruelty case without this being considered to be a prosecution or a persecution of a shochet performing his religious duty. If the carefully elaborated millennial practice of killing under the laws of shechita was not adhered to, the case no longer dealt with the Jewish method of slaughter at all. Letters from members of the Jewish community demonstrated that local sentiment drew the distinction with clarity and precision. "The man Myers was summoned, not because he followed the Jewish method, but because in the discharge of his duties he was unskillful or careless of both. Had he carefully followed the Jewish system, he would not have been prosecuted, nor could the bench have inflicted a fine upon him."[21]

If this truly were the case, one might ask—why did Colam introduce the case to the court as the first time the Jewish method of slaughtering animals was being brought before the magistrates; why underline the essential cruelty of shechita; why follow these limiting remarks about pursuing Myers for departing from the practice, by drawing the magistrates' attention to the ordinary method of slaughtering oxen by stunning or killing them with a poleax?

There is an insurmountable problem with the rhetorical structure and the legal taxonomy of the prosecution's case, even if one accepts that Myers botched the job and the ox was *treif* and not suitable for consumption by observant Jews.[22] The point of reference is the normal method of stunning prior to bleeding the animal. If Myers was engaged in killing the ox using the Jewish method, and his failure was to miss the single cut, then the issue of stunning is a logical irrelevance if compared to the normal method. After having failed the cut, and rendered the animal unfit for Jewish consumption, Myers should have stunned the beast, but this had nothing to do with the use of the poleax in the ordinary English method of slaughter. The rhetorical slippages betray the true framework in which this RSPCA prosecution of the shochet took place. The actual aims of the RSPCA in its campaign against shechita and in favor of what it defined as humane slaughter, of which stunning was an essential element, always emerged from the arguments embodying the true goals of the prosecution.

The confusion is demonstrated by the way in which Colam articulated the facts that he believed demonstrated the ill-treatment of the animal by Myers.

---

21 "The Jewish Method of Slaughtering Cattle," *Liverpool Mercury*, October 8, 1895. (Letter from S. Friedenberg).

22 And ye shall be holy men unto Me; therefore ye shall not eat any flesh that is torn *(treifa / terefah)* of beasts in the field ye shall cast it to the dogs. (Exodus 22:30).

The barrister argued that the ox was cast and that the process took at least six minutes during which time the animal was "writhing in abject fear." This stage of the RSPCA argument replicates assertions in other prosecutions about the process of casting for shechita and repeats positions about animal suffering during the process that other courts had rejected. The next part of the prosecution case went to Myers's lack of skill and the resultant cruelty. The barrister recognized that in the normal course of the Jewish method, once the animal is cast and its neck exposed, the shochet proceeds immediately to the single killing cut. It was alleged that Myers had delayed the process by sharpening his knife while the ox remained in its cast position, prolonging the suffering and that when he had cut the neck he had succeeded in severing only two of the four main blood vessels. Seeing the botched process, another butcher had stepped forward with a poleax to put the ox out of its misery but Myers had stopped him. Only after hearing that an RSPCA inspector was present did Myers allow another butcher to sever the remaining blood vessels and to dispatch the ox with a poleax.

The argument about animal cruelty targets actions outside the rules of shechita. Myers failed to cut the ox's throat in the required manner, so the animal could no longer be one considered to have been killed according to Jewish law. By interfering with the quick dispatch of the animal, Myers would have materially contributed to its suffering, and would be considered to have done so outside the rules of Jewish law and ethical command relating to the treatment of animals. However, Colam could not stop there. He added that as far as the RSPCA case was concerned, he believed that it was cruelty under the terms of the statute not to use the poleax on an animal once its throat had been cut. He called for stunning after cutting as the only way the Jewish method could be lawful, extending his argument beyond the facts of the case and returning to a generalized attack on shechita. The issue of the infliction of a stunning blow after the shochet has cut the animal's throat and its permissibility under *halacha* remains contentious to this day. The RSPCA case was one that at many levels was nothing more than an attempt to outlaw the Jewish method of slaughter.

Evidence from an RSPCA inspector and from Dr. J.F. Ryder, a veterinary inspector for the United States government, both of whom had been present when Myers botched the killing, confirmed Colam's argument. Myers, arriving late and without legal representation, left unchallenged this testimony establishing that he had failed to slit the animal's throat in the manner prescribed for shochetim. For the first time in England, a shochet was convicted for cruelly treating an animal during an attempt to practice shechita.

Myers was fined twenty shillings including costs. The lenient sentence was imposed because this case was the first of its kind brought before the court, but the magistrates indicated that any future prosecution and conviction would receive a much harsher sentence. The court opined that the time spent in the casting process was longer than necessary, and needed to be reduced.[23]

Myers' only statement to the court was to deny that he interfered with the use of the poleax following the botched cutting. He had not expected the trial to take place when it did, and had not presented or prepared any witnesses.[24] He was taken by surprise that the delay had not been granted and his religious faith ignored by the court. He came to court late and unprepared and faced a set of factual accusations that he still vehemently denied. In his twenty-six years as a shochet, he had never failed to kill an animal according to the religious law. In his later interview, he asserted that he had been prepared to cut the ox's neck immediately after it was cast. Colam's assertion that animals killed by the Jewish method should be poleaxed immediately after the cut would result in the animal being unfit for consumption. Myers pointed to Colam's two-faced claim that the RSPCA respected and did not wish to interfere with any religion's tenets, while immediately thereafter asserting the innate cruelty of shechita. Tumpowsky finished his letter to the *JC* in terms that clearly identify the nature of the Birkenhead prosecution. "I have placed before you the statement of a Shochet of world-wide experience, expert in his profession, and upon the other side evidence of prejudiced witnesses, and it is for the Jewish community to decide if they are going to protect the honoured traditions of our race, by appealing against a decision that appears to me to be an error of judgment."[25]

Myers was convicted for inflicting ill-treatment on an animal he was meant to kill using the Jewish method. Despite the often confused, confusing, and conflated set of arguments proposed by Colam, it might be possible to minimize the significance of the 1895 Birkenhead case. From the magistrates' comments, the casting aspect of the slaughtering was not the basis on which the conviction was entered. The botched cutting by Myers, and his attempts to interfere with the dispatching of the ox by poleaxing, provided the grounds for finding animal cruelty. This was not an instance of shechita properly so-called since it must be carried out following strict and unchangeable rules. The court held that Myers failed to slit the animal's throat in the prescribed fashion, with

23 "The Slaughter of "Kosher" Meat: A Question of Religion and Cruelty," *Blackburn Standard*, October 5, 1895.

24 *JC*, "The Birkenhead Decision Upon Shechita."

25 Ibid.

a single cut. Once that had occurred, the animal was no longer being killed according to Jewish law for consumption by observant Jews. Myers' prosecution was the direct result of failing to carry out shechita.

It is evident from the rhetoric of Colam's presentation to the Police Court at Birkenhead that he conceived of the case as one attacking the inherent cruelty of shechita. Myers was convicted for events directly relating to the Jewish method of slaughter, and even if his alleged gross incompetence took events outside the formal rules of killing animals for Jewish consumption, in the minds of the prosecution, and it seems of the magistrates, this was a case criminalizing Jewish slaughter. Seeing the case as one of persecution and not prosecution, becomes a more attractive interpretive strategy once Myers' own post-trial statements are taken into account. He denied all of the alleged facts and insisted that he had killed the ox according to the tenets of Jewish practice, as he had for twenty-six years. The story of the slaughter as narrated by Colam and the prosecution witnesses, according to Myers, was fantasy. Myers paid the fine on October 3.[26]

Support for the idea that this was a case of shechita improperly performed can be found in the reactions of two sources, the RSPCA and Jewish religious authorities in London. The RSPCA was unambiguous. "We unhesitatingly say that the prosecution and conviction were based solely on the manner in which the process was carried out by the offender."[27] On October 14, 1895, following a review of the evidence and the defense offered, the Chief Rabbi wrote to Myers expressing his concerns about the shochet's competency. Myers was summoned to London to be examined by the Chief Rabbi and others to ensure that his authorization to conduct shechita could continue.[28] From the perspective of the Jewish leader, the prosecution and conviction of Myers offered evidence that he had failed to perform the Jewish method of slaughter in a correct fashion. Soon thereafter Myers traveled to London where he explained the facts to the Chief Rabbi, and the *JC*. He complained about the poor state of facilities at the slaughterhouse in Birkenhead for Jewish slaughter. Finally, he "exhibited his expertness at the Whitechapel market before Ecclesiastical Authorities and several of the Shochetim."[29] No problem was identified with his technical skills and qualifications as a shochet.

---

26 Birkenhead Justices: Police Information and Complaints Books, B/13x/142/20, Wirral Archives Service.

27 "News and Notes," *Animal World*, November 1895, 162.

28 Office of the Chief Rabbi, ACC/2805/01/01/43, Letters from July 18th 1895 to April 23rd 1896, Letter 1009 to J. Meyer [*sic*], LMA.

29 "The Birkenhead 'Casting' Case," *JC*, October 25, 1895.

working at Birkenhead were at that time involved in one of the many periodic disputes between shochetim and Shechita Boards that would characterize the legal history of shechita in England.[52] As a result of the crisis, the amount of kosher meat available for English consumption fell dramatically and the Chief Rabbi had to reach an arrangement with his counterpart in Rotterdam for kosher Dutch mutton to be shipped on the morning after shechita to London. At the same time, the Humanitarian League, another animal welfare charitable group, not limited as was the RSPCA by a dominant and strict Christian worldview, mounted another campaign aimed at outlawing shechita.[53]

Opposition to the Jewish method had reached its zenith with the hearings and publication in 1904 of the report of the Parliamentary Committee on the Admiralty that had condemned shechita in no uncertain terms.[54] Campaigns had been waged by the RSPCA and other advocates opposed to shechita over many years. The Liverpool/Birkenhead cases fit into a social, political, and legal context in which anti-shechita activists employed lobbying and prosecution efforts in their attempts to outlaw shechita, or as they often insisted on putting it, banning all modes of slaughter that did not use stunning. They had succeeded in introducing a general set of such rules through the Board of Trade recommendations, but the Jewish communities had trumped their efforts by obtaining a specific exemption for killing carried out by approved and licensed shochetim. The Jewish communities continued to oppose anti-shechita movements at all stages, and to campaign against them and for the preservation of Jewish religious observance.

The period witnessed broader manifestations of antisemitic discourse around the passage of the Aliens Act in 1905.[55] As Glover argues that, "anti-Semitism was part of a wide-ranging and deeply contested racial imaginary, with its own distinctive debates, polemics, idioms, and practices which formed a cultural matrix that allowed the possibility of anti-alien legislation to become thinkable."[56]

52 "The Shechita Board and the Butchers," "Shechita Board: A Manifesto," *JC*, September 25, 1908.

53 Emanuel, *A Century and A Half of Jewish History*, 180; "The Humanitarian League," *JC*, October 23, 1908; Dan Weinbren, "Against All Cruelty: The Humanitarian League," *History Workshop* 38, no. 1 (1994).

54 Minutes of Evidence Taken Before the Committee Appointed by the Admiralty; Report of the Committee Appointed by the Admiralty; "The Jewish Method of Slaughtering," *Manchester Guardian*, July 28, 1904.

55 5 Edw. 7, c.13.

56 David Glover, *Literature, Immigration, and Diaspora in Fin-de-Siècle England: A Cultural History of the 1905 Aliens Act* (Cambridge: Cambridge University Press, 2012), 80.

A brief summary of Joseph Myers' career offers some insight into his prosecution, and serves to call into doubt a conviction based on allegations of failing to carry out shechita consistently with Jewish law. Joseph Myers had a long history as shochet in Liverpool, although it was not without incident. In 1898, following a series of disputes involving various shochetim, butcher shop owners, and the Shechita Board, the Chief Rabbi suspended Myers from his functions. While he was serving his suspension, it was alleged that he continued to act as shochet for some butchers specializing in killing poultry.[30] Eventually Myers was reappointed and continued to practice his profession. He was so skilled that the Shechita Board, with the approval of the London Beth Din, later appointed Myers to serve as porger (*menakker*), a position requiring a particular set of talents in butchery according to the laws of kashrut.[31]

Reading the Birkenhead case as one in the line of anti-shechita prosecutions brought by the RSPCA becomes more plausible once one realizes that the hearing took on Sukkot II. Myers had informed the court of the religious holiday in his request for an adjournment but that had been denied. It appears that he may not even have received notice, and he showed up late, alone, and unprepared on the day of the trial. The prosecution of a shochet for cruelty in what the RSPCA barrister called the first case against the Jewish method of slaughter took place on a Jewish religious holiday. The hallmarks of a persecution targeting the Jewish method of slaughter rather than a judicial proceeding carried out in the best traditions of British justice seem present.

The Royal Society for the Prevention of Cruelty to Animals, in the person of its secretary, John Colam, the Birkenhead prosecutor's father, rejected these assertions of antisemitic motivations. John Colam acted as secretary of the RSPCA for forty-five years.[32] Speaking to the correspondent of the *JC* during a visit to the Deptford abattoir to investigate new methods of humane killing, Colam denied targeting the Jewish method.[33] The *JC* correspondent reported:

> On my remarking that the prosecution looked very much like an attempt to proscribe the Jewish method of slaughter, he assured me that the Royal Society for the Prevention of Cruelty to Animals had not the slightest intention of interfering, by proceedings under 12 and 13 Victoria, ch. 92,

30 Shechita Board Minute Book, January1, January 3, February 10, February 20, March 13, 1898 and January 12, 1904, LRO.

31 Ibid., December 30, 1901; January 12, 1902; August 2, 1904.

32 "Obituary: Mr. John Colam," *Times*, May 30, 1910.

33 "Humanity in the Slaughterhouse," November 29, 1895.

> with the religious customs of the Jewish community. He had always discountenanced any action that would be open to such a construction of the Society's intention.

He distanced himself from counsel's (his son R.F. Colam) assertion at the trial that the Jewish method was cruel, saying that this argument did not form part of his formal instructions from the RSPCA. He added that had a motion been made in open court, the RSPCA would have been more than happy to acquiesce to an adjournment for the Jewish holiday.

Casual readers of the press might have few doubts as to the core questions of the case, or as to its appropriate characterization and popular connotations. This was a prosecution of the Jewish method of slaughter and its inherent cruelty.[34] The *JC* reproduced the local news account of the trial under the headline "The Jewish Method of Slaughtering Cattle: Prosecution at Birkenhead."[35] For the non-Jewish reading public, the case confirmed the broad understanding that shechita was inherently cruel, even for those who attacked the brutality of slaughterhouses generally.[36]

Because the case took place on a Jewish holy day and Myers had arrived late and unprepared, he had been unable to mount a full defense against the charges, in law or on the facts. His alleged incompetence as a shochet had been reproduced as fact in the press coverage. He had demonstrated the weakness of this part of the claim by practicing his skills before the assembled Jewish experts in London subsequent to the trial, and his many years as a successful and in demand shochet in Liverpool belied to some extent the assertions of gross carelessness at the heart of the RSPCA case. The unchallenged narrative of incompetence and the inherent cruelty of shechita remained on the public record as the RSPCA managed to obtain a conviction and to assert its rhetorical monopoly over issues of humane slaughter in this case. For the first time, a shochet had been convicted of animal cruelty in relation to shechita.

The Second Birkenhead Shechita Prosecution: The Case Against Abraham Friedner

The RSPCA campaign in the Liverpool area continued a year and a half after the case against Joseph Myers. In January 1897, the RSPCA brought another

34 "The Slaughter of 'Kosher' Meat: A Question of Religion and Cruelty," *Weekly Standard and Express* (Blackburn), October 5, 1985.

35 October 11, 1895.

36 "The Jewish Method of Slaughtering Cattle," *Liverpool Mercury*, October 10, 1895.

charge of animal cruelty against Edward Newton and Abraham Friedner.[37] It was alleged that, on December 18, 1896, after having killed three animals according to the "Jewish fashion," Friedner walked away as the fourth, a red bull, was cast. The bull got up, was cast a second time, got up again, and had to be driven into a pen with other animals to "cool down." A half hour later another Jewish butcher killed the bull. The accusation was not a direct attack on the Jewish method of slaughter, since nothing was made of the killing of the first three animals. These were allegations about Friedner's incompetence in letting the bull escape at least twice, and then failing to kill it immediately, leaving the animal in a distressed and tortured state.

The case proceeded in front of Magistrate Craven, who had previously convicted Joseph Myers, and Magistrate Peter McMahon. Local solicitor J.T. Thompson appeared for the RSPCA. Newton does not appear to have been represented by counsel, but W.H. Quilliam presented the case for Friedner. W.H. Abdullah Quilliam was the most famous British convert to Islam in Victorian times.[38] His second son, W.H. Billal, also practiced as a solicitor.[39]

Friedner was described as "the chief Jewish butcher at the lairages," and was a shochet of some experience. Born in Austria, Abraham Israel Friedner became a naturalized British subject on July 19, 1894.[40] His naturalization document describes him as working as a "shochet or examiner of meat for Jews." By 1911, he was also a Hebrew teacher in Liverpool.[41] He had first been authorized to act as a shochet by Dr. Nathan Adler and had worked in that capacity for over twenty years at the time of the Birkenhead prosecution.[42] He was later a key player in a series of industrial disputes among Liverpool's shochetim, butcher shop owners, and the Shechita Board.[43] The prosecution presented evidence that Friedner was in charge of the shochetim at the abattoir and that "No other butcher, however, could touch a beast unless Friedner gave him the order." The RSPCA witnesses each testified that Friedner had simply walked away and refused to kill the animal.

---

37 "The Jewish Method of Slaughtering Cattle: Prosecution at Birkenhead," *Liverpool Mercury*, January 15, 1897. "Prosecution of a Liverpool Shochet," *JC*, January 22, 1897.

38 Jamie Gilham, *Loyal Enemies: British Converts to Islam, 1850–1950* (London: Hurst, 2014).

39 Jamie Gilham, "'Upholding the Banner of Islam': British Converts to Islam and the Liverpool Muslim Institute, c.1887–1908," *Immigrants and Minorities* 33, no. 1 (2015).

40 Naturalisation Certificate, A7948, HO 334/21/7948, National Archives.

41 Census Returns of England and Wales, 1911, RG 14, Piece 1663, National Archives.

42 "A Shochet's Evidence," *JC*, February 10, 1911.

43 "The Kosher Meat Trade," *JC*, December 25, 1908.

Friedner's co-defendant, Newton, gave evidence. He was in charge of slaughtering for the Swift Beef Company and had engaged Friedner to kill fifteen cattle for shipment to London. Friedner did not complete the job. Quilliam argued that there was no evidence of intentional cruelty by Friedner, who had believed that someone else would kill the animal in question. Friedner asserted that as shochet he had nothing to do with the process of casting, and that he had informed those present that he could kill only three animals before he was needed elsewhere. The casting took place when he was absent and without his knowledge or control.[44] The court found that no case had been made out against Newton, but found Friedner guilty of cruelty and entered a fine of ten shillings and costs. Friedner paid his fine on January 22. A week earlier, on January 16, Newton received six pounds six shillings in costs from the complainant RSPCA Inspector Pocock.[45]

A shochet had again been found guilty of an animal cruelty offense in relation to the practice of shechita. The prosecution did not allege any cruelty by Friedner in the killing of the animal, since the issues in the case had arisen from the fact that Friedner had allegedly refused to kill the beast. The witnesses proffered by the RSPCA described the usual course of killing using the Jewish method, as Friedner had done for the first three animals that day, as quick and over within ten seconds. Friedner was widely regarded as the best shochet at Birkenhead and the testimony established his flawless record. The alleged cruelty involved the difficulties associated with the failed castings and the escape of the animal. The RSPCA witnesses testified they had not heard Friedner order the casting of the bull, but instead they had only seen him walk away. The facts, or RSPCA allegations, of cruel treatment relate not to the Jewish method of killing but to the failure of Friedner to proceed to the killing of the animal by the Jewish method. Friedner was later condemned by the Chief Rabbi on these same grounds. His failure to proceed with the killing of the animal resulted in suffering and the inhumane treatment was attributable to him. His refusal or failure to act was contrary to Jewish law, and his authority as a shochet was suspended.[46]

---

44 "Prosecution of a Liverpool Shochet."

45 Birkenhead Justices: Police Information and Complaint Books, B/13x/142/21, Case Number 5 (Friedner) and Number 64 (Newton), Wirral Archives Service.

46 Office of the Chief Rabbi, ACC/2805/01/01/45; Minister Melbourne Minute Book and Letter Book, December 1896– March 31, 1897, Letter 3832, A. I. Friedner, January 18, 1897, LMA.

The reactions of the religious authorities to Friedner's failures to comply with the demands of Jewish law do not explain the perspective of the prosecuting officials. The process of casting, whatever its factual specificities, is clearly associated with killing under shechita because the animal must be prepared and placed in a position that allows the single killing cut by the shochet. The liability imposed on Friedner is clearly based, in the mind of the court, on his position as the chief Jewish butcher at the abattoir. The bull was being prepared for killing by the Jewish method; no one could kill an animal without Friedner's say so. Newton who had overall charge of the killing and butchering process was found to have no case to answer. The court assumed and imposed liability on the shochet because it understood that the entire process from casting to killing was Jewish, and Friedner was the only Jew involved. The entire culpability for the cruelty inflicted on the animal came down on Friedner, either because he was seen as being in charge of the failed process of casting, despite the absence of any evidence that he had ordered it, or because he had omitted to kill the animal quickly and in accordance with the practice of shechita.

This is arguably not a shechita case, and could instead be characterized as one in which the laws of shechita were not followed by the shochet. Ironically, given the history of RSPCA prosecutions, it was a non-shechita case where the shochet was held liable for not performing the quick, single cut killing of the bull. Friedner was found guilty in his supervisory capacity for the failures of the casting process, despite the significant lack of evidence that he was in charge of the preparations. Given the wording of the statute relating to accessorial liability and the complexities involved in proving that someone was guilty of inflicting or causing cruelty and ill-treatment by doing nothing, in other words, liability by omission, the conviction of Abraham Friedner at Birkenhead appears to have set new jurisprudential standards in animal cruelty law and in the prosecution of shechita. Non-Jewish newspaper coverage of the case left little room as to the proper lay understanding of the trial—"The Jewish Method of Slaughtering Cattle: Prosecution at Birkenhead."[47] In the space of fourteen months, two shochetim had appeared before Magistrate Craven charged with cruelty offenses and each had been convicted. There is an argument about each case to show that neither involved the process of killing by the Jewish method, without stunning, as the core of the criminal charge, but that does little to dispel the broader implications of these as cases in which Jews

47 *Liverpool Mercury*, January 15, 1897.

were convicted on animal cruelty charges related directly in the minds of the RSPCA prosecutors and the magistrates, to the cruelty of shechita.

The Final Birkenhead Prosecution: "Kosher Butcher Fined"

What appears to have been the final RSPCA prosecution in England took place at Birkenhead several years later. In this case, a butcher's assistant, W. M'Laughlin, was charged with animal cruelty for brutally kicking a cow and gouging its eye in an effort to bring the animal to its feet to be recast, following the original casting in which it had fallen on its stomach rather than its side as required.[48] M'Laughlin asserted in his defense that he had caused no pain to the bullock but the prosecution presented evidence that after slaughtering, it was found that the animal's eye was lacerated. This was entered as evidence that the bullock had "suffered great pain." The defense presented an argument grounded firmly in Jewish law. "So strict were the Jewish tenets, even had the animal been beaten with a stick on its way to the slaughter the Jewish butcher would decline to kill it, and the flesh of any animal that had been ill-treated during the process of slaughter could not be sold as Kosher beef."[49]

M'Laughlin was convicted and fined forty shillings and costs or fourteen days' imprisonment. For the third time at Birkenhead, someone involved in the slaughter of a bullock by the Jewish method had been found guilty of animal cruelty. The same arguments raised in relation to Joseph Myers and Abraham Friedner can be asserted in this case. This was a casting case rather than a shechita case strictly speaking and Jewish law does not prescribe a method or process of casting. As the defense argued, the tenets of Jewish law in such instances prohibit the acts ascribed to M'Laughlin. While no method of casting is required, what is compulsory is that the animal to be killed is not in any way injured or damaged, and that it is treated with kindness and mercy. A lacerated eye and several kicks to the side of the bullock could mean that the animal could not be killed by the Jewish method because such an animal would not be fit for consumption by observant Jews. This 1908 case is finally distinguishable from the Myers and Friedner convictions because the person convicted was not a shochet but a slaughterhouse butcher's assistant, and judging from his name, he was not Jewish. He was not a shochet, authorized to kill under Jewish law.

---

48 "Cruelty at the Lairages: Kosher Butcher Fined," *Manchester Guardian*, September 26, 1908.

49 Ibid.

Despite this, for many, and in concrete ways, the case was understood as involving the slaughter of a bullock by the Jewish method. The bullock was being prepared for killing by the shochet and the cruelty arose during the process of placing the animal in a position for the administration of the killing cut. As far as newspaper readers were concerned, this was yet another case involving the Jewish method of slaughter. The accused is described as a kosher butcher when he was a (non-Jewish) butcher's assistant. Yet the supervising shochet was not charged. Unlike the Friedner case, where much of the prosecution argument was grounded in the overall supervisory jurisdiction of the shochet, only the person who physically mistreated the bullock was accused. This is perhaps attributable to the correctness of the defense's assertion that such actions as alleged (and proven) would place the animal outside the category of one that could be killed for the purposes of kashrut and no shochet would reasonably participate in such acts or permit them to occur. This is not a shechita case, but once again it is more precisely a "not shechita" case. The legal proceeding, as with the earlier incident involving Friedner, had nothing to do with the actual single cutting motion or the absence of pre-stunning. But, as the newspaper headline made clear to its readers, there was cruelty at the lairages, and that cruelty was directly related to kosher killing.

"Kosher Butcher Fined": The Context of Early Twentieth-Century Shechita

In the autumn of 1908, kosher butchers occupied a significant place in the public's mind. The Liverpool Corporation had recently passed an order requiring the pre-stunning of all animals slaughtered in the Corporation's abattoirs. It had inserted an exception for shechita, exempting Jewish butchers from the requirement, after long discussions involving the Shechita Board and its communications with the office of the Chief Rabbi.[50] The Liverpool model adopted the position taken by the Board of Trade in a circular to local councils, who had jurisdiction over the day-to-day operation of abattoirs. As a result of the intervention of the Board of Deputies, the Jewish method of slaughter had been made a specific exception to the pre-stunning rule.[51]

The application of the exception for shochetim applied only to those who had been approved and licensed by local Ecclesiastical Authorities. The butchers

50 "The Shechita Board and the Butchers," *JC*, October 23, 1908; 296 KAS 1/1, Shechita Board Minute Book, 1904, July 27, 1904 and September 4, 1905, LRO.

51 Emanuel, *A Century and A Half of Jewish History*, 177.

Part of the discursive world creating and reflecting anti-Jewish agitation was the continuing attack on shechita and the insistence on stunning as the only humane slaughtering method. Debates about limiting immigration and the general social and political agitation around the influx of "aliens," many of whom were Jews fleeing pogroms, state violence and exclusion, and extreme poverty in eastern Europe, grabbed the attention of Jewish authorities, government officials and politicians, and antisemites in equal measure. Antisemitism was part of the zeitgeist. Anti-shechita agitation allowed some Jew-haters to focus their attention on what they saw as the inherent barbarity and cruelty of the practice. The influx of unclean, and perhaps even blood-loving, eastern European Jews only exacerbated the danger in the minds of those who insisted that passage of the Aliens Act was a national priority.

At the same time, Jewish communities continued to struggle with intra-communal disputes about regulatory and governance monopolies exercised by Shechita Boards over the licensing and approval mechanisms on which the Jewish exemption had come to rely. Into this complex mix, had come three cases in the northwest of the country in which criminal convictions had been entered against those involved in shechita. But as far as the records indicate, the later nineteenth- and early twentieth-century cases at Birkenhead marked the final use of animal cruelty prosecutions by opponents of the Jewish method of slaughtering animals for human consumption. The battleground shifted to the political, legislative, and regulatory realms in England. But this does not end the legal history of attempts to criminalize shechita.

CHAPTER 6

Criminalizing Shechita Down Under: The Sydney Case

The Jews of Sydney and Shechita, 1882

In September 1882, the anti-cruelty movement's battle against the Jewish method of slaughter came to Sydney. The Animals' Protection Society, which had changed its name from the Society for the Prevention of Cruelty, charged the shochet, Philip Philipstein (Philippstein) with cruelly ill-treating a bullock at the Glebe Island abattoir. As with prosecutions of shochetim in the United Kingdom, the accusation of cruelty and ill-treatment in the first instance related to the practice of casting. Counsel for the Society insisted that pithing animals, destroying their brain and spinal cord with a sharp implement, was preferable because it was an instantaneous method of killing and that the Jewish mode was "an unnecessarily cruel way of slaughtering animals."[1] The Australian case mirrored the United Kingdom cases, with its two-track emphasis on the particulars of casting the animal in preparation for slaughter by the shochet and its true, or primary, attention on the act of killing the animal without stunning.

This attack came against a Sydney Jewish community that had established significant roots in the city. There had been Jewish convicts among the First Fleet that had landed in Botany Bay in 1789, beginning the process of white settlement of the country. Other Jewish convicts arrived with subsequent ships. Many of these convicts found their freedom and later ordinary Jewish settlers joined them as part of the broader expansion of the British Empire in the Antipodes, particularly during the Gold Rush period of the middle of the nineteenth century.[2]

---

1 "Jewish Mode of Slaughtering Cattle," *Jewish Herald* (Melbourne, VIC), September 22, 1882.

2 Suzanne D. Rutland, *Edge of the Diaspora: Two Centuries of Jewish Settlement in Australia* (Sydney: Collins, 1988); Hilary Rubenstein, *Chosen: The Jews in Australia* (Sydney: Allen & Unwin, 1987).

The original white settlers and particularly the military and colonial administrators during the earliest years of settlement had brought with them a univocal and established Anglican Church.[3] Years of struggle had loosened the grip of established Anglicanism, especially in favor of a Roman Catholic Church, to serve the spiritual needs oflarge numbers of convicts, and then freed settlers, from Ireland. By 1836, disestablishment had occurred and all Christian denominations were placed on an equal footing.[4] Jews appear to have benefitted from "full civil and political rights: they acquired British nationality, voted at elections, held commissions in the local militia, were elected to municipal offices and were appointed justices of the peace."[5]

The situation in which the Jews of New South Wales emerged as emancipated British subjects arose from local conditions, and the decisions of local colonial officials, rather than as the result of formal legislative intervention. As Getzler points out, the Chief Justice, Alfred Stephen, adopted the position that the Plantation Act of 1740 that permitted Jews to take an oath of public office, omitting phrases such as "on the true faith of a Christian" that were part of the standard declaration, was a sufficient legal basis for fuller emancipation. The Plantation Act had allowed many dissentient "Protestants and others," including Jews, in the colonies to acquire British nationality, own property, and exercise economic and political rights. Stephen chose to ignore the subsequent Justices Oath Act of 1766 that had re-imposed the "true faith of a Christian" form.[6]

Well before their brethren in England enjoyed unfettered access to public office and civic and political rights, the Jews of Sydney could and did participate as equal citizens with their Christian counterparts. The one area in which they experienced inequality was in formal public support for their religious institutions such as burial grounds, synagogues, and schools. New South Wales governments granted requests for support by individual legislation, by way of special permissions, rather than under generally applicable statutes.[7] The

3 Matthew Allen, "The Politics of the Pew: Faith, Liberty, and Authority in a Sydney Church in 1828," *Journal of Religious History* (2016), doi: 10.1111/1467-9809.12407.

4 An Act to promote the building of churches and chapels and to provide for the maintenance of ministers of religion in New South Wales, 7 Wm. IV, c. 3 (1836).

5 Israel Getzler, *Neither Toleration nor Favour: The Australian Chapter of Jewish Emancipation* (Melbourne, VIC: Melbourne University Press, 1970), 11.

6 7 Geo. III, c. 9.

7 An Act to make provision for the sale and disposition and for the application of the proceeds of certain Lands and Hereditaments in the City of Sydney held upon Trusts for a Jewish

formal exclusion of the Jewish religion from a statutory regime meant to ensure the "practical advancement of the Christian religion," stood as a roadblock to full equality for Jews in New South Wales for some time.[8]

Jews in Sydney were nonetheless broadly equal. Individuals such as Maurice Alexander, Morris Adler, Henry Emmanuel Cohen, John Jacob Cohen, Samuel Cohen, Charles Collins, Samuel Emmanuel, Solomon Herbert Hyam, and Samuel Aaron Joseph, all served in the New South Wales legislature between 1859 and 1901. John Jacob Cohen later became Speaker of the Legislative Assembly. Cohen, like his namesake, Henry Emmanuel Cohen, after a career at the Bar, served on the Bench. In law and in politics, New South Wales Jews served with distinction, on all sides of the divides of colonial politics, as they continue to do to the present day.[9]

Jews were free to practice their religion without hindrance.[10] Like Jews in other parts of Australia, Sydney's Jewish community fell organizationally under the aegis of the Chief Rabbi in London, who served not just British Jewry, but the Jews of the British colonies as well.[11] Not only did the Chief Rabbi act as the supreme ecclesiastical authority for Sydney Jewry, but the London Board of Deputies took steps to preserve their rights and status. In 1847, the Board of Deputies appealed successfully to the Colonial Office to obtain an exemption from work on Jewish Holy Days for Jewish prisoners.[12]

A system of Jewish worship and of Jewish religious governance was in place in Australia and in Sydney at the time of the shechita prosecution. The question of kashrut had been raised in early colonial days, as it had been in North America. The status and functions of those identifying themselves as shochetim had been discussed within the Australian Jewish communities.[13] Like their colonial American counterparts, Jews in Australia found it difficult, if not impossible, in the early days of the colonies, to adhere strictly

---

Synagogue and School and for the appointment of Trustees for such purposes and to declare Trusts of Lands and Premises appropriated for a Jewish Synagogue and School, 36 Vict. (1872).

8 Getzler, *Neither Toleration nor Favour*, 12 et seq.

9 Geoffrey Brahm Levey and Philip Mendes, eds., *Jews and Australian Politics* (Brighton: Sussex Academic Press, 2004).

10 W.D. Rubinstein, *The Jews in Australia* (Melbourne, VIC: Australasian Educa, 1986).

11 Jeremy I. Pfeffer, *'From One End of the Earth to the Other': The London Bet Din, 1805–1855, and the Jewish Convicts Transported to Australia* (Brighton: Sussex Academic Press, 2008).

12 Emanuel, *A Century and A Half of Jewish History*, 53.

13 Pfeffer, *'From One End of the Earth to the Other,'* 288–290.

and unwaveringly to the dietary laws of their religion. This was exacerbated, ironically, by the early emancipation and equality granted to Jews, who were generally accepted within the wider communities of Australia. This led to a high degree of social integration and a move away from some aspects of identitarian existence, including keeping kosher. There was a general religious laxity and the regular consumption of non-kosher meat. The absence of qualified shochetim, and for Jews not living in the major cities, the huge distances that needed to be travelled to obtain kosher meat, meant that important aspects of the American Jewish experience were directly mirrored for Australian Jews in the mid- to late-nineteenth-century period.[14]

In cities such as Sydney, the situation would gradually improve. Increasing populations meant that there were more Jews, and, as in both North America and the United Kingdom, the type of Jewish migrants meant that there were more Jews who wished to practice their religion and observe Jewish law. There was a sufficient number of Jews to support the practice of shechita and to pay for the process of providing kosher meat. In 1837, the first synagogue had been created in rented premises and by 1844, a purpose-built Sydney synagogue, Australia's first such building, was erected following a subscription campaign that drew support from Jews and from well-wishers within the Christian community.[15] The York Street synagogue was replaced in 1878 by Sydney's landmark Great Synagogue, on the corner of Park and Elizabeth Streets. At the time of the prosecution of Philip Philipstein in 1882, Sydney Jewry had established itself as a vibrant and, after some years of divisions and tensions, a unified part of the city. Religious life, including shechita, operated under the watchful eye of the Sydney rabbi, the Reverend Alexander Barnard (Bernard) Davis.

Born in London in 1828, Rev. A. B. Davis took up his first post as leader of a Jewish congregation in 1854 in Kingston, Jamaica. He arrived in Sydney to take charge of the York Street synagogue in August 1862. As he had done in Jamaica, Davis managed to heal the rift in the Sydney Jewish community, reuniting the York Street congregation and the dissenters who had split to set up a congregation at the Macquarie Street synagogue. In 1878, he consecrated the Great Synagogue, from where he retired and became emeritus in 1903, until his death in Sydney in 1913. Throughout his time in Sydney he was known as a religious innovator, founding the Jewish Sabbath School, admitting girls to

---

14 Rutland, *Edge of the Diaspora*, 72–73.

15 Suzanne D. Rutland, *The Jews in Australia* (Cambridge: Cambridge University Press, 2005).

religious classes, and conducting classes for them in the synagogue itself. He created a mixed choir in the synagogue, believed to have been the first of its type in the British Empire.[16]

Philip Philipstein, the shochet accused of animal cruelty in 1882, would pronounce the eulogy at the Davis graveside.[17] Philipstein, who died in 1932, acted as one of Sydney's shochetim for many years. He held a variety of posts within the Great Synagogue, his obituary notice describing him as "Rev. P. Philipstein, a former minister of the Great Synagogue."[18] He was an assistant minister to Rabbi Davis.[19] The German-born Philipstein had been trained by one of that country's leading teachers in the practice of shechita. He worked as a shochet in the United Kingdom, having received his certificate of competence countersigned by the Chief Rabbi, Nathan Adler. He had served his congregation there as Cantor and Teacher. He was appointed to Sydney in 1880. During his time in Sydney he was active as the Jewish prison chaplain, officiating at Long Bay jail each Friday, and occasionally travelling to the prisons at Goulburn, Bathurst, and Parramatta to minister to Jewish prisoners. His interests extended to charity work aimed at the rehabilitation and reintegration of released prisoners and he was a committed Zionist.[20] He occupied important positions in the synagogue, acting as *hazan*, as well as serving as shochet under Rabbi Davis, who appointed him a member of the Sydney Beth Din. When Davis travelled to England, Philipstein deputized for him during his absence.[21]

While Philipstein as shochet stood alone before the court as the individual charged with acts of animal cruelty, throughout the entire story of the Sydney shechita case, the two leaders of the Great Synagogue became the embodiment of the struggle for Jewish religious freedom and for the dignity of the Sydney Jewish community.

---

16 G. F. J. Bergman, "Davis, Alexander Barnard (1828–1913)," *Australian Dictionary of Biography* National Centre of Biography, Australian National University, http://adb.anu.edu.au /biography/davis-alexander-barnard-3379/text5113.

17 "The Late Rev. A. B. Davis: Impressive Funeral," *Sydney Morning Herald*, December 18, 1913.

18 "Obituary: Rev. P. Philipstein," *Sydney Morning Herald*, November 17, 1932.

19 Hilary Rubenstein, *The Jews in Australia: A Thematic History, Volume One, 1788–1945* (Port Melbourne, VIC: William Heinemann Australia, 1991), 532.

20 Ibid., 551.

21 "Rev. P. Philippstein," *Hebrew Standard* (Sydney), August 6, 1926.

Animals, Cruelty, and Law in Australia's Victorian Era

The 1882 prosecution of Philip Philipstein for animal cruelty at the Glebe abattoir did not arise out of thin air. The relative ease with which the Jews of Australia had achieved legal emancipation, and the assimilation of many Jews in larger Australian population centers, did not mean that antisemitism was absent from Australian society.[22] Sydney was essentially a large English city located in a British colony. The mores and values of Victorian England, including antisemitism in various forms, permeated the upper echelons of the New South Wales ruling class, many of whom at this time were English-born.[23] Federation was still almost twenty years away in 1882, and Sydney was the capital of the British colony of New South Wales. The civilizing mission of Victorian Christianity was as much a part of Sydney society as it was in London, especially as those in the colony fought to throw off, or to bury, the convict stain at the heart of the experience of New South Wales' history as a penal colony.

Just as part of the Christian worldview of social and moral improvement in England revolved around the elimination of cruelty, especially in the treatment of animals, Sydney society concerned itself with the same moral and social questions.[24] Anglophile sensibilities, Victorian worldviews, and an abhorrence of the violence, drunkenness, and squalor of the lower classes combined to make the Sydney animal advocacy groups a politically powerful force in the colony. The first Society for the Prevention of Cruelty to Animals was founded in Victoria in 1871, followed by Tasmania in 1872, New South Wales in 1873, South Australia in 1875, Queensland in 1883, and Western Australia in 1892.[25]

The legal regime of animal protection in the Australian colonies reflected that in effect under English law. In 1850, New South Wales introduced the

---

22 Sol Encel, "Antisemitism and Prejudice in Australia," *Patterns of Prejudice* 23, no. 1 (1989).

23 In 1851, the *Argus*, the leading newspaper in Melbourne had re-published an article from an English newspaper condemning shechita, but nothing appears to have come of it. "The Jewish Code of Slaughtering Animals for Food," May 9, 1851. This was the era of general agitation around the question of Jewish equality in Victoria. "Meeting of the Jews of Melbourne," February 23, 1854; "Legislative Council," October 30, 1852. The *Argus* would never be known for its philosemitism. Jim Usher, *The Argus: The Life and Death of a Newspaper* (Melbourne, VIC: Australian Scholarly Publishing, 2007); Muriel Porter, *The Argus: The Life and Death of a Great Melbourne Newspaper (1846–1957)* (Melbourne, VIC: RMIT Press, 2003).

24 K.D. Baker, "Vivisection debate in nineteenth century Great Britain: a muted echo in colonial and early post-colonial Australia," *Australian Veterinarian* 76, no. 10 (1998).

25 "Our History," http://www.rspca.org.au.

core statute, An Act for the more effectual prevention of Cruelty to Animals.[26] Like the English act, the New South Wales statute targeted those who would "cruelly beat ill-treat over-drive abuse or torture or cause or procure to be cruelly beaten ill-treated over-driven abused or tortured" animals for punishment by way of a fine not to exceed five pounds. (§1). Public hygiene concerns led to the creation and introduction of a system for the regulation of slaughterhouses along models similar to those being gradually developed and introduced in Mother England.[27] In 1850, the public abattoir at Glebe Island was established as the only place within the city of Sydney where animals could be legally slaughtered.[28] In order to ensure hygiene and to situate these facilities as far away from local habitation as possible, other restrictions were legislatively imposed.[29] Public health measures concerning slaughterhouses and the disposal of unhealthy animals supplemented the existing regulatory regime.[30]

The SPCA, later the Animals Protection Society of New South Wales (APS), came into existence within the context of this regulatory and penal regime covering the operation of the Glebe Island abattoir and the broader prevention and punishment of the ill-treatment of animals. From its creation, the Society had the support of the highest echelons of Sydney society. Among the founding members were Sir Alfred Stephen and John Fairfax. "Their ideas and sentiments often governed the actions of many, not only in the exciting and absorbing questions of social and political life, but also in the long-neglected work of humanity, regarding the recognition of the rights of our dumb and defenceless animals."[31]

John Fairfax had emigrated from England in 1838 and purchased the *Sydney Herald* in 1841, changing its name to the *Sydney Morning Herald* the next year. He became not just the publisher of Sydney's most influential

26 14 Vict. c. 40 (1850).

27 Marc Trabsky, *Institutionalising The Public Abattoir in Nineteenth Century Colonial Society*, 40 Australian Feminist Law Journal 169 (2014).

28 An Act for the establishment of a Public Abattoir or place for the slaughtering of Cattle for the City of Sydney, 14 Vict., c. 36 (1850); Phil McManus, "Feeding a growing city: The Glebe Island Abattoir and the provision of meat for Sydney," *Rural Society* 11, no. 3 (2001).

29 An Act to amend the Law relating to the slaughtering of sheep Cattle and other Animals within certain Towns in the Colony of New South Wales and for the prevention and removal of certain Nuisances therein, 14 Vict., c. 30 (1850).

30 An Act to amend the Laws for the slaughtering of Cattle and to secure the immediate destruction of animals dying of disease, 15 Vict., c. 13 (1851).

31 F. Montagu Rothery, "Animals Protection Society of N.S.W. Incorporated," *Carcoar Chronicle* (NSW), March 5, 1897.

newspaper, but an important man in the world of business in the New South Wales capital. He was, like many of his class in this era, heavily involved in charity, leading the Young Men's Christian Association and cofounding the Congregational Church. He "was well known for his tolerance at a time when sectarian feeling ran high."[32] He served on the Council of Education and in the partly appointed upper chamber, the New South Wales Legislative Council.

Sir Alfred Stephen's reading of the Plantation Act had laid the legal groundwork for the political emancipation of New South Wales Jews. Born in St. Kitts, Stephen moved to London where he was called to the Bar at Lincoln's Inn. In 1825, he sailed to Van Diemen's Land (Tasmania), and almost immediately upon his arrival in the colony was appointed solicitor general and crown solicitor. In 1833, he was named the colony's Attorney General, and maintained a career in government and business, as well as a lucrative private practice. In 1839, he moved to Sydney to take up a temporary position on the Bench. The position was made permanent and Stephen became a prominent jurist and the principal drafter of many key pieces of colonial legislation in New South Wales. In 1845, he became the third Chief Justice of the colony. Like many of his class at the time, Stephen was active in his church and in charitable offices. He was a member of the Australian Anglican Diocesan Committee, and he was the key actor in the creation of St. Paul's College as the affiliated Anglican college at the University of Sydney. While he would sever all ties with the Anglican Church following a dispute, he worked tirelessly for the Temperance Society and the Female Refuge Society, and remained an active advocate of temperance as a cure to many social ills, especially in relation to the violence inflicted on women, throughout his life. Stephen was a man of his age and a progressive. In 1886, he had introduced legislation permitting a wife to sue for divorce on several grounds, including her husband's habitual drunkenness.[33] Although the bill failed to pass into law, Stephen persisted in his campaigns for social betterment and served on a number of charities for several years, including taking an active role in the Society for the Prevention of Cruelty to Animals.[34]

32 J. O. Fairfax, "Fairfax, John (1804–1877)," *Australian Dictionary of Biography* National Centre of Biography, Australian National University, http://adb.anu.edu.au/biography/fairfax-john-3493/text5357.

33 Sir Alfred Stephen, *Australian Divorce Bills: The Objections Raised to Them, Religious and Social, Considered* (Sydney, 1888).

34 John M. Bennett, *Sir Alfred John Stephen: Third Chief Justice of New South Wales* (Sydney: Federation Press, 2009); *Portraits of The Chief Justices of New South Wales, 1824–1977* (Sydney: John Ferguson, 1977), 19–21; Martha Rutledge, "Stephen, Sir Alfred

Fairfax and Stephen embodied the ideas and policies of social reform and Christian good works at the heart of the Victorian era, and at the foundational core of the Society for the Prevention of Cruelty. Fairfax used the pages of the *Sydney Morning Herald* to promote the ethical and moral vision of the SPCA. The networks of information, ideology, and solidarity between the metropolitan center and the colonial outposts circulated through the only available mass medium, the newspaper.[35] Stephen on more than one occasion intervened in the Legislature to promote the animal protection agenda. The SPCA engaged in a familiar two-pronged assault against animal cruelty. It sought the introduction and enforcement of legislative prohibitions on ill-treatment on the one hand, and it sought to inculcate the moral values of humaneness and "the powerful agency of enlightened views."[36]

The re-named Animals' Protection Society fought for the introduction of compulsory moral education on issues of animal cruelty in the New South Wales school curriculum.[37] They lobbied the New South Wales government to adopt changes to the teaching of pupils to include "object lessons and instructions in kindness to animals."[38] They investigated and reported to high-ranking governmental officials concerns about the cruelties inflicted on animals during the slaughtering process at the Glebe Island abattoir.[39] Stephen attempted to introduce legislation to improve and expand the protections against animal cruelty.[40]

By the time the Animals' Protection Society sought to prosecute the shochet for animal cruelty in 1882, the legal framework and the socio-political context had long been established. Cruelty against animals was a social evil, one condemned in legislation and in a complex set of statutory regulations meant to reduce suffering in the slaughterhouse at Glebe Island. Leading members of the Sydney establishment had placed the humane treatment of animals on the social agenda and on the law reform table. The time to attack shechita in

---

(1802–1894)," *Australian Dictionary of Biography* National Centre of Biography, Australian National University, http://adb.anu.edu.au/biography/stephen-sir-alfred-1291/text7645.

35 Simon Potter, "Communication and Integration: The British and Dominions Press and the British World, c. 1876–1914," *Journal of Imperial and Commonwealth History* 31, no. 2 (2003).

36 Rothery, "Animals Protection Society of N.S.W. Incorporated."

37 "Prevention of Cruelty to Animals," *Sydney Mail and Advertiser*, September 7, 1878.

38 "Animals' Protection Society," *Sydney Morning Herald*, September 4, 1880.

39 "News of the Day," *Sydney Morning Herald*, April 19, 1879.

40 "In the Legislative Council," *Sydney Morning Herald*, September 26, 1878.

the name of (Christian) humanity had arrived, and had been bubbling away elsewhere in the antipodean colonies for some time.

Criminalizing Shechita in Australia: Rumblings in Adelaide

The free colony of South Australia was founded by Act of Parliament in 1834.[41] Unlike New South Wales, it was not established as a penal colony and members of all faiths were permitted to immigrate to populate the area. By 1846, at least fifty Jewish settlers could be found in the city of Adelaide. Two years later, in 1848, the community had reached sufficient size and financial standing that the Adelaide synagogue, just off Rundle Street in the center of the city, was erected.[42] Abraham Tobias Boas, a native of Amsterdam, served as rabbi and shochet from 1870 until 1918.[43] By 1876, Morris Lyon Marks, Emanuel Solomon, and Judah Moss Solomon had served as members of the colonial legislature in South Australia. Judah Moss Solomon was Mayor of Adelaide in 1869–1870. Jews had quickly established an important social and political presence in the city.

Six years before the Sydney prosecution, shechita and allegations of cruelty associated with the Jewish method of slaughter had entered public debate in Adelaide. In 1876, the newly founded South Australian Society for the Prevention of Cruelty to Animals had sent an inspector to the Adelaide slaughterhouse to witness and report on Jewish slaughter.[44] As in Sydney, conditions at the local abattoirs were a continuing target of significant complaints, about animal ill-treatment, and public hygiene.[45] South Australia did not have a separate regime of anti-animal cruelty legislation. Acts of mistreating animals were prohibited by Section 67 of the general Police Act.[46]

41 An Act to empower His Majesty to erect South Australia into a British Province or Provinces and to provide for the Colonisation and Government thereof, 5 Will. IV, c. 95 (1834).

42 Rutland, *Edge of the Diaspora*, 39–40; Rubinstein, *Chosen*, 22–23, 104–106.

43 Louise Rosenberg, "Boas, Abraham Tobias (1842–1923)," *Australian Dictionary of Biography* National Centre of Biography, Australian National University, http://adb.anu.edu.au/biography/boas-abraham-tobias-5277/text8897; "Rev. A.T. Boas: A Pioneer Jewish Minister," *Australian Jewish History Society Journal* 7, no. 2 (1972).

44 "Heads of Intelligence," *South Australian Advertiser* (Adelaide), January 1, 1876; Wallace B. Budd, *Hear the Other Side: A History of the Royal Society for the Prevention of Cruelty to Animals in South Australia 1875–1988* (Adelaide, SA: RSPCA, 1988).

45 "Central Board of Health," *South Australian Advertiser*; January 27, 1876; "The City Cattle Market and Slaughterhouse," *South Australian Chronicle and Weekly Mail* (Adelaide), April 24, 1880; "The City Slaughter-Houses," *South Australian Register* (Adelaide), November 21, 1879.

46 An Act to consolidate and amend the Law relating to the Police in South Australia, 33 Vict., c. 15.

It was an offense, subject to a fine not exceeding five pounds or imprisonment, with or without hard labor for a period not exceeding two calendar months, to "wantonly and cruelly beat, ill-treat, overdrive, overload, abuse or torture" any animal or procure such ill-treatment.

In the spring of 1876, Mr. W. Mofflin, under the pseudonym "Humanity" wrote to the local press on the subject "Cruelty to Animals."[47] In his letter, Mofflin complained about "a system of most barbarous cruelty" to which he wished to draw the attention of the police and the SPCA. "Had I not witnessed the revolting operation I could not have believed that so cruel a system of murder was being carried on from day to day." Mofflin offered a detailed description of casting operating at the Adelaide slaughterhouse. He finished his screed by asserting that he had no objection to Jewish dietary laws, but he wished that they could be adhered to "without continually subjecting dumb animals to be tortured to death." At the next meeting of the City Council, one member, Stacy, drew his colleagues' attention to the Mofflin letter, and read it to the assembled representatives. He informed them that upon his own investigations with the Inspector of Slaughterhouses, he found that the letter's allegations that "the slaughtering of cattle for the Jewish community was carried out in a barbarously cruel manner" were correct.[48]

The discourse in both Mofflin's correspondence and in Councilor Stacy's public intervention is familiar. There is nothing, they were careful to state, in their concerns that expresses a desire to, as Stacy put it, "interfere with any section of the religious community," but at the same time he could only regret that "such abominable cruelty" was carried out on a regular basis at the city slaughterhouse. The concern was not with the Jewish religion or with Jews; it was not antisemitic in its origins or motives. The focus was on the suffering caused to the animals and the accompanying cruelty and barbarity. Jews were not cruel or barbarous, but their religious practices were. Stacy was encouraged and supported by his colleagues who seconded his call for all killing in the slaughterhouse to be carried out under the system of pithing. The Council was insisting that shechita be banned because it was barbarous and cruel and it passed a motion that the Inspector of Slaughterhouses be summoned to appear before the Health Committee to report on the Jewish method of killing.

47 *South Australian Register*, April 1, 1876.

48 "Municipal Corporations," *South Australian Register*, April 4, 1876; "At the City Council," *South Australian Advertiser*, April 4, 1876. The matter was addressed at Council and Health Committee meetings on April 3, April 18, and May 1, 1876.

Mofflin continued his letter-writing campaign alleging acts of cruelty "in the slaughtering of cattle for the use of the Jews."[49] Part of Mofflin's objections related to the preparatory practice of casting, but his principal problem was with "the old system of murder." Shechita was, without qualification, not just murder, but "a barbarous system of cruelty." The step between the expression of outrage informed by a moral and ethical worldview about how animals should be treated, and unmitigated expressions of vile antisemitism, with references to blood lust, murder, and cruelty was a small one. Mofflin's correspondence was followed up by a number of other letters to the editors of local papers, all of which supported his attack on the extreme brutality of the Jewish method of slaughtering. Because they were anonymous or pseudonymous, a common practice at the time, the newspapers did not deem them suitable for publication on such a charged issue.[50] The aura of antisemitic outrage and agitation caused by the anti-shechita campaign raised significant concerns for the publishers and editors of Adelaide's newspapers. Mofflin continued his onslaught against the SPCA for failing to put a stop to the practice, and against shechita, the unnecessarily and barbarously cruel method of killing, and finally against the Jews of Adelaide for not exercising their authority to put an end to the cruelty.[51]

Mofflin's campaign, supported by important voices on the Council, had its impact. The Inspector of Slaughterhouses was summoned to the Health Committee of the Adelaide Corporation.[52] Immediately following the formal statutory establishment of Adelaide as a Municipal Corporation, the city had been given a monopoly over the slaughtering of all cattle within its boundaries at a public slaughterhouse.[53] That monopoly had carried with it an indirect financial benefit to the city, since the Slaughterhouse Inspector was given the right to levy fees for his remuneration on every animal killed in the public abattoir.[54] The Committee heard evidence from the inspector "in regard to the

---

49 "Cruelty to Animals," *South Australian Register*, April 8, 1876.

50 "The Jewish Mode of Slaughtering," *South Australian Register*, April 12, 1876.

51 "Cruelty to Animals," *South Australian Register*, April 17, 1876.

52 Since its statutory creation, the City of Adelaide and its Council had been known as the Corporation. An Act to constitute a Municipal Corporation for the City of Adelaide, 4 Vict., c. 4 (1840).

53 An Act to regulate the Slaughtering, and prevent the Stealing of Cattle, 4 Vict., c. 5, §2 (1840). The monopoly and regulatory regime remained largely unchanged throughout the relevant period. An Act to amend the Laws relating to the Corporation of the City of Adelaide, and also to regulate the Slaughtering of Sheep, Calves, Swine, and Goats in the said City, and to prevent certain nuisances therein, 18–19 Vict., c. 18 (1855–1856).

54 An Ordinance to Authorize the Levying of Fees on the Slaughtering of Cattle in South Australia, 7–8 Vict., c. 2 (1844).

manner of slaughtering cattle under the Jewish regulations."[55] The Committee members held their decision in abeyance pending a visit to the abattoir to witness shechita for themselves. An inspector of the South Australian SPCA twice visited the Adelaide Corporation slaughterhouse to witness the shochet killing cattle. He reported to the secretary of the SPCA that the procedure was the same on both occasions and that the animal had been killed within five minutes of its entry into the slaughterhouse and within two minutes of casting. He then witnessed five animals killed after stunning. "I do not now presume to offer an opinion as to the degree of cruelty in either or both cases, believing that it can only be determined by scientific investigation."[56]

The SPCA forwarded the inspector's report to the local press to allay fears that the killing of cattle for members of the Jewish community was innately and obviously cruel. Although the inspector did not proffer an opinion, it is clear that he did not condemn the Jewish method as cruel, or crueler than the practice of stunning or pithing, nor did he invoke the provisions of the Police Act against the shochet.

The SPCA Inspector, Hubert Arnold, penned a swift and stern reply to Mofflin's persistent attacks on the organization, his report, and the Jewish mode of slaughter. He did not wish to further engage with Mofflin, who "appears to have a peculiar plan of drawing inferences, and is certainly very liberal in his insinuations."[57] Arnold repeated that his reports had been based on his personal attendance at the slaughterhouse and observing the two methods of killing cattle. He reasserted the basic outcome of his observations. He was unable and unwilling to state any conclusion about the relative cruelty of the methods of killing practiced at the abattoir without scientific evidence that he was not in a position to provide. Arnold did offer two statements that belie this reluctance to give an opinion. He concluded his correspondence by noting that "by the last mail the opinions of some eminent men on the mode of slaughtering have reached the colony, and the system adopted by the 'Jewish butcher' is spoken of as the easiest, quickest, and best."

In addition to its substantive support of shechita, Arnold's statement is yet another manifestation that information about the Jewish method of slaughter circulated among animal protection groups, and others, as part of an international network for the transfer of knowledge. He wrote in reply to

55 "Health Committee's Report: City Slaughterhouses – Cruelty to Animals," *South Australian Register*, April 20, 1876.

56 *South Australian Chronicle and Weekly*, April 15, 1876.

57 "Cruelty to Animals," *South Australian Register*, April 20, 1876.

Mofflin's attack on Jewish murder and cruelty that "a case was tried years ago in London, when Serjeant Ballantyne was retained for the defence of the 'Jewish butcher,' and obtained a verdict in his favour."

The 1855 case against Yankoff Cohen in London was known to SPCA officials, and now the general public, in Adelaide.[58] The SPCA inspector recognized and informed the broader public that the idea of a scientific debate on the issue existed and doubt persisted on the relative and comparative merits of slaughtering methods. As it stood for the South Australian SPCA, the Jewish method of slaughter, after public scrutiny, had passed legal muster.

Mofflin persisted, assailing the process of casting, and pointing out that the shochet was morally responsible for the cruelty inflicted at this stage, even if he was not personally involved in preparing the cattle for slaughter. He attacked the SPCA inspector's report and insisted that because the matter was in the hands of the public authorities, all steps necessary had to be taken to end the practice: "If the practice of slaughtering cattle in the way described by me for the Jews, has as stated by Mr. Arnold, existed for hundreds of years, the more necessity that something should at once be done to put a stop to it." [59]

The visit by members of Council to the local slaughterhouse to witness the Jewish mode of slaughtering and "the ordinary mode of slaughtering practiced in that establishment" took place near the end of April, and no further action was taken in relation to shechita.[60] The question of the Jewish method of slaughter had petered out, and now left the arena of public discourse and political debate. Nine years later, one pseudonymous letter to the editor of the Adelaide *Express and Telegraph* complained that the Jewish method of slaughtering cattle was "horribly cruel," and called for SPCA intervention, but no further discussion of the question appears to have taken place in Adelaide until some years later.[61]

The SPCA in South Australia held its first annual meeting on May 30, 1876, one month after the public controversy surrounding shechita and the role of its inspector in the issue. In addition to noting "the growing sympathy manifested by the public" for its work, the Secretary's Report highlighted the effort expended in inspecting and condemning the unsanitary conditions at the slaughterhouse, and the neglect of animals kept there. "As to the cruelty

---

58 "Alleged cruelty of Jewish Butchers," *North Australian, Ipswich and General Advertiser* (Ipswich, QLD), February 5, 1856.

59 "Cruelty to Animals," *South Australian Register*, April 22, 1876.

60 "Municipal Councils: The Mayor's Report," *Adelaide Observer*, May 6, 1876.

61 "Sympathy," August 18, 1885.

of the Jewish mode of slaughtering the Inspector had reported that the allegations were not supported."[62]

For the next few years, the Jewish method of slaughtering animals for human consumption appears to have left the public consciousness of the people of Adelaide. This was in large part at least attributable to the decline of the SPCA, described in its own history as "moribund" between 1890 and 1904.[63] The resurgence in interest in the practice of shechita and attacks on the Jewish method of slaughter came about when the SPCA began a new and revived phase of activity. In 1904, there was a return to some of the ugliness that had surrounded the 1876 investigations. Another pseudonymous letter to the *Adelaide Advertiser* attacked the "great cruelty" of the Jewish method, again focusing on the casting process.[64] While Humanitas did pretend to focus on the casting aspect of shechita, the letter was a thinly veiled attack on the killing of animals without stunning. The letter specifically refers to the Admiralty Committee report, condemning the Jewish method and all killing without stunning. As the information and knowledge about shechita and the unrelenting attack from the British Parliamentary Committee circulated throughout the empire, the writer furthered the antisemitic tone by noting that since Jews were known for ignoring many aspects of their religious rights or obligations, interference with one more in the form of the rules of shechita would be of little import.

Humanitas was quick to shift the burden in the matter to "our Jewish friends." If they could only be persuaded to abandon their practice, any "suspicion or slur of cruelty" would be removed. The letter marked an unfortunate return to the discourse of Jewish perfidy that has consistently surrounded attempts to use the machinery of state intervention to ban shechita. The religious practice of observant Jews was defined as irredeemably cruel and inhumane. Killing with stunning was the only humane way to slaughter animals. Jewish persistence in the practice of following the holy injunction not to damage or injure the flesh of an animal before it is killed was the source of suspicions and overt charges of cruelty. To rid themselves of such slurs they needed only to abandon a fundamental tenet of their practice, followed for thousands of years, and yield in the name of Christian humanity.

---

62 "Society for the Prevention of Cruelty to Animals," *South Australian Advertiser*, May 31, 1877.

63 Budd, *Hear the Other Side*, 26.

64 "Slaughtering Cattle," September 3, 1904.

The secretary of the SPCA, Janie Ennis, replied to Humanitas and informed readers that the Society was concerned about these issues. Ennis was secretary of the South Australian SPCA from 1905 to 1911. She was responsible in large part for the revival of the organization at that time and led its campaign for the introduction of a "Kindness to Animals" course in the state's schools. Ennis took a particular interest in the question of humane slaughter and represented the South Australian group in 1924 at the centenary celebrations in London for the RSPCA, where once again, the subject of humane slaughter featured prominently on the agenda.[65]

The SPCA had consulted the "Jewish Butcher," Mr. Kither, whose presence insured that the slaughter of cattle was "as good as it can ever be under existing custom."[66] The SPCA representative repeated and echoed the ongoing concerns about the nature and practice of casting and the suffering involved. Ennis echoed Humanitas' reference to the Admiralty Committee's Report, and insisted that the finding that stunning was the only humane method had to be imposed in all public slaughterhouses. Such action was required to allow "dumb animals to be treated as the Creator meant them to be." The *Advertiser* had performed a great public service by allowing Humanitas to bring the question of the Jewish method of slaughter again to the public's attention. The SPCA would not remain silent when confronted with a condition of animals that was "one of the greatest wrongs done on God's earth."

SPCA efforts in behalf of the poorest of God's creatures resulted in attempts to introduce specific legislation in South Australia dealing with animal cruelty, taking the existing provisions of the Police Act and expanding them into a legislative framework comparable to that in the UK and some other Australian jurisdictions. Correspondence in the local press pressed home the inhumanity of the Jewish method and called for the compulsory introduction of stunning in the state's abattoirs as part of the proposed animal cruelty legislation.[67] Some members of the legislature echoed Humanitas' argument and insisted that "the Jewish people might be fairly asked to adopt a modification of the present method" by including stunning in the Jewish method.[68] The compromise expected of the Jewish community of Adelaide was, in the name of humanity, to abandon their millennial religious belief and practice. The level of misunderstanding, possible ignorance, and perhaps willful blindness, not to

65 Budd, *Hear the Other Side*, 24, 77–78, 86.

66 "Slaughtering Cattle," *Adelaide Advertiser*, September 5, 1904.

67 "Humanity," *Adelaide Advertiser*, September 18, 1908.

68 "Cruelty to Animals Bill," *Adelaide Advertiser*, September 30, 1908.

mention the possibility of a not too subtle anti-Jewish bias, persisted in welfare, legal, and governmental circles from the 1876 debates to the revival of anti-shechita proposals in the first decade of the twentieth century.

In 1908, the legislature of South Australia passed the state's first comprehensive anti-cruelty statute. An Act for the Prevention of Cruelty to Animals repealed the old provisions of the Police Act and instituted a new regime of animal protection.[69] In addition to the prohibitions on the ill-treatment of animals that form the core of this and similar legislation in the Anglophone world, the statute regulated slaughtering and prohibited any act in relation to animal slaughter that subjected any "animal to unnecessary pain or suffering." (§ 4 (g)). But this was not a victory for the South Australian SPCA, Janie Ennis, or Humanitas. Section 4 (4) of the same statute provided:

> Nothing in this Act shall render unlawful the slaughtering of any animal in any manner which may be necessary to comply with the requirements of the Jewish or other religion.

Attempts to prohibit shechita in South Australia through the mechanism of the criminalizing provisions of animal cruelty laws, or the subsidiary means of local slaughterhouse regulation, had been beaten off by the Jewish community and those in the broader society who placed a value of religious freedom and tolerance. The exemption amendment had been moved by the government and supported with eloquence by an important member of the state legislature, Sir John Downer. Downer was the founder of a South Australian political dynasty, a prominent member of the Adelaide Bar, and served as Attorney General from 1881 to 1884. He played a key role in the process of Australia's move to federation and was elected as one of South Australia's six senators to the first federal Parliament. After leaving national politics, Downer was elected to the South Australian Legislative Council where he served until his death in 1915.[70]

Downer spoke in favor of the amendment introducing a Jewish exemption after having spent many hours meeting with members of the Adelaide Jewish community. He told his legislative colleagues that he had read the Admiralty Committee Report. He dismissed it with a strong dose of irony, condemning the "false humanitarianism" of those who attacked the alleged cruelty of

69 8 Edw. VII, c. 956 (1908).

70 Peter Bartlett, "Downer, Sir John William (1843–1915)," *Australian Dictionary of Biography* National Centre of Biography, Australian National University, http://adb.anu.edu.au/biography/downer-sir-john-william-6007/text10261.

shechita in the British report. He added that the logical result of accepting the notion of Jewish barbarity in killing animals according to religious law would have been either to once again banish the Jews from England or "compelling them to become vegetarians."[71] During subsequent debates some members objected to the exemption on the grounds that no cruelty should be justified by religion, while others asserted that the inclusion of the clause, as an exemption from the operation of a statute dealing with animal cruelty, might be seen as legislative approbation of the idea that the Jewish method of slaughtering was cruel, and could only be saved by way of a special provision. Downer insisted that the clause had the clear support of the Jewish community, and no such inference about shechita being cruel could or should be drawn from a measure insisted upon by the Jewish community itself.[72] Shechita debates in Adelaide were ended by Section 4(4) of the Prevention of Cruelty to Animals Act, 1908.

One further element puts the Adelaide investigation of 1876 and subsequent SPCA interventions in the early twentieth century, as well as the New South Wales prosecution of Philipstein, into a different or somewhat more complex social, political, and theological light. The shochet, the Rev. Mr. Boas, rabbi of the Jewish congregation in Adelaide under attack from Mofflin, was himself an active member of the SPCA. In his intervention before the Council on April 4, Stacy had highlighted the fact that "the moving spirit" or "one of the principal agents" behind the recent founding of the SPCA "took an important part in the slaughtering."[73] The secretary of the SPCA who had sent Arnold's report on shechita to the press was none other than Abraham Abrahams, one time Treasurer and President of the Adelaide Hebrew Congregation. His obituary declared that he was "a consistent Jew."[74]

A businessman, Abrahams was active in local philanthropic ventures, including the Home for Incurables, the Society of Arts, the Public Library, and the Art Gallery. He was one of the oldest Freemasons in South Australia, and served as a Justice of the Peace. He was identified with the SPCA, "an organization which was practically founded by him." For Boas and Abrahams, the ethical and divine command to act in ways consistent with *rachamim*, mercy and compassion, as for the Josephs and Meldola de Sola in Montreal, led them to obey the laws of kashrut and to play key roles in the SPCA without any sense

71 "In the Council," *Adelaide Register*, September 23, 1908.

72 "Cruelty to Animals Bill," *Adelaide Advertiser*, October 15, 1908.

73 "Adelaide Corporation: Cruelty to Animals," *Express and Telegraph*, April 4, 1876.

74 "Obituary: Death of Mr. Abraham Abrahams," *South Australian Register*, April 12, 1892; "Death of Mr. Abraham Abrahams," *Adelaide Observer*, April 9, 1892.

of a moral contradiction. In its earliest stages, the South Australian SPCA had abjured the narrow Christian biases of its English counterpart, and adopted a more ecumenical humanitarianism as its founding and driving ethos.

Part of the failure of the SPCA in Adelaide to continue its work in the period between 1890 and 1904 was the slow withdrawal and finally the death of the energetic Abraham Abrahams. When it revived in the early twentieth century, in order to insure its flourishing and its acceptability to South Australian society, the SPCA featured the Governor as its President, and the Chief Justice, and the Lord Bishop of Adelaide, and a number of clergy and elected politicians among its directors. Among those on its committee was Rabbi A.T. Boas, and the mayor of the city, Lewis (later Sir) Cohen.[75] Judah Moses Solomon had served as the first Jewish mayor of Adelaide in 1869–1870, and Cohen was the second Jew to lead the city between 1889–1891, 1901–1904, and 1909–1911. He was Lord Mayor from 1921–1923. While the Christian clergy and the Protestant elite of South Australian society featured prominently in the newly emergent SPCA, Jews figured as they had from the beginning.

Jewish involvement in anti-cruelty movements was at the time well known. Louis Gompertz in England was one of the founders of animal protection in that country. The Joseph brothers in Montreal were key members of the Canadian Society. As representatives, theological and legal, of the defendants in the anti-shechita prosecutions from England, to Scotland, to Sydney, to Nova Scotia, would consistently demonstrate, Judaism is a religion of kindness, and one that in the killing of animals for consumption gives primacy to the well-being of the animal. It is hardly surprising given these basic theological tenets, that Abrahams and Boas should have been active advocates for the prevention of cruelty to animals, and adherents and practitioners of kashrut and shechita.

The contradictions between animal anti-cruelty advocacy and Jews arose from the Christianization of the movement in Victorian England, and the influence of a theologically infused ideology among many proponents. Attacks on shechita became grounded in antisemitic ideas and imaginings of Jewish blood lust, cruelty, and the murder invoked by Mofflin. In the years following Abrahams' death in 1892, Christianization began to take hold at the heart of the Adelaide SPCA. Ennis' 1904 intervention, in the name of the SPCA, in echoing and publicizing calls for the end of slaughter without stunning, is anti-Jewish in its intended effects, and is sprinkled with phrases and terminology that indicate that ending such cruel practices is a task imposed by (a Christian) God.

75 Budd, *Hear the Other Side*, 20–30.

Those understandings and that ideology formed the basis of the prosecution of Philip Philipstein in Sydney six years after the original Adelaide investigations.

The Glebe Island Abattoir Case of 1882: A Sydney Shochet in the Dock

The trial of Philip Philipstein took place on Monday, September 11 at the Water Police Court in Balmain, the court with jurisdiction over the nearby Glebe Island abattoir.[76] Mr. Robey, probably Arthur John Robey, a local solicitor,[77] acted for the prosecution and David Levy, an active member of the Great Synagogue congregation, engaged by the Jewish community, appeared for Philip Philipstein.[78] The stipendiary magistrate, James Buchanan, presided. Buchanan followed his early job as Goldfields Commissioner with a stint as a member of the New South Wales legislature, and had a long career on the magistrate's bench, finishing as magistrate at the Sydney Central Police Court.[79]

The case began in a recognizable fashion. The APS inspector testified about events at the slaughterhouse on August 25, commencing with a detailed description of the process of casting the bullock, throwing it violently on a cement floor, and raising it into a position for its throat to be cut by the shochet. The inspector testified that the process had taken fifteen minutes during which, in his opinion, the beast showed signs of terror and pain. He further opined that the alternative mode of slaughter, by pithing or some other form of blow to the head, was instantaneous and would cause no pain to the animal. He informed the court that the APS had received many complaints about conditions at Glebe Island and that he had been instructed to take prosecutorial action. There was no indication that the complaints or his instructions related specifically to the Jewish method of slaughter, but an inference to that effect might be drawn from his evidence comparing the cruel and painful Jewish mode and the painless pithing alternative.

Levy introduced his case by informing the court that the issues before it had already been "mooted in England some years ago," a reference to the Cohen prosecution at the Mansion House in London, a case which was known

76 "Jewish Mode of Slaughtering Cattle," *Sydney Morning Herald*, September 12, 1882. The quotations are taken from this source.

77 "Deaths," *Sydney Morning Herald*, September 8, 1886.

78 "The Late Mr. D.L. Levy," *Jewish Herald* (Melbourne: VIC), December 15, 1893.

79 http://www.parliament.nsw.gov.au/prod/parlment/members.nsf/1fb6ebed995667c2ca256ea100825164/7fa7d483e5f34ee3ca256cb8000ac8a8?OpenDocument.

in Australia, and of which the Jewish community, and Philipstein, who had received his certification as shochet under the auspices of the Chief Rabbi in London, would have been aware. The defense argument was that Philipstein was trained in shechita; he was a representative of Sydney's Jewish community when he performed the act under summons; and the Jewish method of slaughter was not inhumane or cruel. The legal case mounted by the defense was one that involved the relevant statutory language of intentionally inflicting cruelty, as well as an implicit argument based in religious freedom. The framing of the Sydney case by the defense was identical to that adopted in England, and which would be invoked later in Scotland, Canada, and the United States, as Jewish communities fought to keep their religious practice out of the grasp of criminalization by animal welfare advocates.

On cross-examination, the inspector admitted that he had never before witnessed the Jewish method of slaughter and that he was unaware that shechita had been carried out without incident or complaint in Sydney for the past twenty-eight years. He persisted in classifying what he had seen as unnecessary and gross cruelty. Under further cross-examination, he confessed that he had no knowledge of the science of pain, and that he knew nothing of the pain involved in the pithing alternative he himself had suggested. Levy dismissed his evidence by accusing him of woeful ignorance and of bringing "trumpery charges" without any real scientifically based knowledge of the pain suffered by animals during any method of slaughter.

The evidence of the inspector was confirmed by the gatekeeper of the Glebe Island slaughterhouse, who placed particular emphasis on casting, and suggested that better technology in the form of an iron ring to pull the animal's throat existed. Like the inspector, he persisted in his overall claim that the process from casting to cutting the bullock's throat was cruel from start to finish because it was a "brutal manner" of killing. When he was cross-examined about the cruelty of pithing, and the pain suffered by an animal killed with this method, Levy was interrupted by the magistrate who stated to the accused's solicitor that he was not helping his case. For the magistrate, the case was limited to the cruelty (or not) of shechita. Levy's attempt to demonstrate that pithing might be cruel did not go to the issue before the court. As far as the magistrate was concerned, the line of questioning about pithing in comparison to the Jewish mode was a factual and legal irrelevance. "What has that to do with this case? It won't influence me one bit."

A. B. Davis was called for the defense. The spiritual leader of Sydney's Jews had not been present at the slaughterhouse, but, he testified, the method described was that followed by Jews in places such as London and Paris; scientists had determined it to be humane; it was a practice commanded by Jewish law. He expressed his surprise that a shochet like Mr. Philipstein should now appear before a court charged with animal cruelty. He added that he and Mr. Philipstein had discussed the method of casting and would consider adopting the London ring method for future killings in Sydney. The magistrate dismissed Davis' testimony in its entirety as irrelevant. "Mr. Buchanan said he had admitted the evidence in deference to Mr. Davis and the high position he occupied, but he did not think it was admissible. He was pretty sure that generalities of that kind were not admissible. What he (Mr. Buchanan) had to try was that identical cast. If Mr. Levy had any evidence to offer with regard to it he should be glad to hear it."

The magistrate limited the case to a narrow understanding of the facts, and of the prosecution's burden. Evidence of comparative cruelty in relation to pithing was irrelevant because the issue before him was simply whether the case against the shochet as charged evidenced cruelty prohibited by law. Expert evidence about Jewish law and Jewish slaughtering practice was not relevant because this was not a case about the Jewish method of slaughter or Jewish religious obligation. It was limited to the way a single bullock had been treated at the Glebe Island slaughterhouse on a particular day as charged in the summons.

Levy called Frederick Oatley, Inspector of Abattoirs, and an inspector at Glebe for twenty years, who began his response to questions from the defense lawyer by stating that in his opinion, the process of casting followed in the Jewish method of slaughter was "rather cruel." The beast was roped, thrown on the floor, and then raised so its throat could be cut, all of which Oatley considered to be crueler than pithing. He concluded by stating that despite his opinion, it would be otherwise impossible to place the animal in a position where its throat could be cut. He added that he "would never take action, because he thought it was a religious observance, which he would not interfere with."

The magistrate interrupted the defense case and brought the trial to a surprising, and less than fully satisfactory, conclusion. He focused as he had throughout on a limiting understanding of the case, yet the judgment rendered still managed to escape the narrow confines of positive law, as shechita cases had done and would do again. Buchanan pointed out that there was a specific mens rea or intentional requirement under the New South Wales statute. If an act were done for a lawful purpose, or had been legalized by custom, then it could not become the wanton and willful infliction of cruelty required by the legislation. Slaughter for human consumption has been, and continues to be, a lawful purpose under animal cruelty legislation. When killing is carried out for that end, it is extremely difficult to make it fall under the legislative anti-cruelty regime. "There could not be a doubt that the act complained of was done for a lawful purpose, and, according to the testimony adduced for the defendant, was legalized by custom, it having been done for years and years; and under those circumstances, the act was not in the category of offences for which punishment could be inflicted."

There is sufficient ambiguity in Buchanan's statement to prevent us from knowing if in this case the lawful purpose was human consumption or religious obligation. There is a gap that cannot be bridged and prevents us from knowing if the "years and years" of customary practice refer to the evidence that shechita had been carried out in Sydney at the abattoir for almost thirty years, or if this is a reference to Rabbi Davis' testimony about Jewish tradition and law. Whatever the case, the magistrate found that Philip Philipstein had not committed an offense under the statute.

The magistrate still felt the need to express his personal opinion, attacking the process of casting as described by the witnesses as "a frightful act of cruelty." He did not mention anything about killing without stunning, or killing by the shochet's single cut. His condemnation of the cruelty was limited to casting, and he was neither the first nor last judge to express such a view, in which he would be joined by members of the Jewish communities where shechita had been and would be in the dock. The rabbi in his curtailed testimony had indicated that both he and Philipstein planned to change the casting procedure at Glebe. But whatever the magistrate's individual views of the casting process, he followed the legal rules incumbent upon him pursuant to his finding that the purpose of killing the bullock was a lawful one, and one legalized by custom. The case was dismissed.

It is not implausible to suggest that the Balmain magistrate found an inherent cruelty in the Jewish method of slaughter and acquitted Philip Philipstein because the cruelty of the Jewish method had been legalized by custom. Had the magistrate applied a narrower legal positivism, and simply found the act to have been done without the requisite wanton cruelty, a broader affirmation of shechita as a humane mode of slaughter might have come out this case. This was a goal of Levy and Rev. Davis who each pointed out and emphasized the existence of scientific evidence to support their broad claim that shechita was not cruel. But this was not a line invoked by Buchanan.

He found that the purpose was a lawful one, and the practice one legalized by custom, leaving him free to condemn at least part of the practice legalized by custom as cruel, but not technically unlawful. The dismissal of the summons against Philip Philipstein was a narrow victory, and a victory is better than a defeat, but the grounds of the dismissal offered less than a resounding vindication of Jewish religious practice under New South Wales law.

Other accounts of the case give a flavor of the mitigated victory for the Jewish community. In Sydney, the *Evening News* headlined its coverage "The Jewish Mode of Slaughtering" and reported that the magistrate found the practice to be "horribly cruel" and a "brutal and barbarous custom." The paper reported that the magistrate had instructed Philip Philipstein "to be less cruel."[80] The *Sydney Mail and Advertiser* gave prominent space to the court's denunciation of the brutality to which much testimony had been dedicated.[81] There can be little doubt that in the mind of the reporter, the case was a narrow technical escape for Philipstein and a condemnation of barbarous Jewish practice. This idea of a brutal Jewish practice legalized by custom was repeated in some rural and interstate coverage of the case.[82]

Magistrate Buchanan's denunciation of the cruelty, and the ambiguity with which he reluctantly dismissed the summons against the shochet set the framework for as series of attacks against shechita and against Jews and Judaism in the pages of the local press. The Sydney *Protestant Standard* led the charge in language and style typical of the sectarianism that informed some inter-denominational relations at the time in Sydney, but that had rarely manifested itself in such virulent antisemitism. After reminding its readers of the perfidy of "the Papal Church," the paper turned its attention to the Glebe Island case. The writer objected to the way in which Papists and now surprisingly, the Jews, hid evil under the mantle of religion. "We certainly gave that denomination (the Jewish community) credit for something better and more liberal than in these days to shelter a palpable abuse—a shocking cruelty under the altar of tradition and custom."[83] The target of the outrage was the practice of shechita. The Protestant writer was clear. "That 'religion' must certainly be very revolting of which a necessary part is to protract needlessly for fifteen minutes the most

80 September 11, 1882.

81 "The Week," September 16, 1882.

82 "The Jewish Mode of Slaughtering," *Australian Town and Country Journal* (Sydney), September 16, 1882; "Humane Slaughtering," *Bendigo Advertiser* (VIC), September 14, 1882.

83 September 16, 1882.

awful tortures with which a poor animal could be afflicted." Such a reaction is hardly unexpected given the rhetorical framing of the coverage of the case and the magistrate's begrudging dismissal of the charge against Philip Philipstein, and his freewheeling, albeit non-binding, denunciation of the brutal cruelty testified to in his court. Whatever the motivations of the APS in bringing the prosecution, there can be little doubt that decisions to pursue shechita under animal cruelty legislation create situations in which social and theological antisemitic stereotypes about Jewish cruelty and bloodlust can be given a juridically justified public airing. The Protestant vehicle gave vent to the worst of these prejudices and turned its aim on A.B. Davis. Using his testimony that he had not witnessed an act of shechita for some time, the paper advised him to experience the cruelty at first hand. "Perhaps his contact with the world, and his intercourse with his flock, and his administration of holy things, and his family relationships to children have humanized him so as to counteract that hardening of tradition, and he may allow mercy in future to triumph over custom."

Jewish custom is clearly neither humane nor merciful. Animals are "hideously tortured for religion's sake—just as for many a century Jews themselves were tortured." The freedom granted to Jews in Australia should and must lead them to grant similar freedom to the beast "and spare their religion the odium of such cruelty." The *Protestant Standard* reflected and embodied hundreds of years of Protestant antisemitism. Jews and the Jewish religion suffer from the odium of cruelty and the hideous custom of the Jewish method of slaughter. Humanity and a true holy faith might allow Rev. Davis to see the error of his ways and denounce the customs of torture, blood, and cruelty inherent in his religion. None of his own testimony about the religious obligations that inform shechita or the evidence of the humane nature of shechita attested to by non-Jewish scientific evidence, or centuries of theological demonstrations of the Jews' holy obligation to kindness and mercy could counter-balance, or even be mentioned, in this anti-Jewish diatribe from the Protestant press.

Rabbi Davis responded to the wave of antisemitic discourse set loose by the Balmain decision, but he was prevented from immediately entering the fray by his "intercourse with his flock." In 1882, Rosh Hashanah fell on September 14, Yom Kippur on September 23, and Sukkot on September 28. The wave of anti-Jewish abuse in Sydney after the failed prosecution of Philip Philipstein circulated around the Jewish Holy Days. Davis found time to write a public letter on September 18. Davis tempered his response with a statement of regret if the animal had suffered unnecessarily. At the trial, he had already expressed concern about the casting process and apparatus used, and had urged the adoption of the London method. He had testified that he had discussed the question with Philip Philipstein and they had agreed to proceed that way in the future. He would attend at the abattoir with all shochetim to introduce new casting procedures. He went into the broader theological (and legal) issues about shechita, explaining publicly that all Jews and shochetim in particular are taught under religious law to have particular concern and respect for the animal to be killed. He highlighted for the uninformed among the readers that not only was inflicting unnecessary pain on the animal contrary to Jewish teachings, but injuring the animal in the way described by some of the prosecution witnesses would make the animal unsuitable for consumption by observant Jews.[84] Other readers of the Jewish press in neighboring Victoria combatted the antisemitic assault unleashed by the Philipstein case by emphasizing the moral core of Judaism and the essential idea that cruelty and brutality diminish the fundamental practices of mercy and compassion at the heart of Jewish life.[85] Jews were compelled to insist that they did not enjoy or practice cruelty, and to urge a broader social understanding among non-Jews about the centrality of *rachamim* to the ethical norms of Jewish existence.

Davis' letter mustered and repeated all of these familiar arguments invoked in favor of the humane nature of the Jewish method of slaughter. As a general theological matter, cruelty was not a part of the Jewish religion. As a more particular matter of the religious laws and obligations relating to shechita, not only was cruelty against the tenets of the faith, but inflicting injury on the animal before it was killed was a direct violation of the purpose and function of the divine injunctions ruling the mode of killing. He accepted that work remained to be done to ensure that the preparatory steps relating to casting were equally humane. The Sydney story is little different from the earlier tales from London and Manchester, and from later events in Liverpool, Aberdeen, and Halifax. The reaction to the case in Sydney, as evidenced by the statements in the *Protestant Standard* and other public responses to the shechita case, is perhaps unique in the vehement and public antisemitism to which it gave rise.

Humanitas, a popular pseudonym for anti-animal cruelty authors of letters to the editor in nineteenth-century Australia, asserted that no sect "should be allowed to carry out a rule in total opposition to the humane laws of this country."[86] The evidence of the case established "a gross piece of cruelty" and

84 "The Jewish Mode of Slaughtering Animals," *Sydney Morning Herald*, September 19, 1882.

85 "Cruelty to Animals," *Jewish Herald*, October 6, 1882; "The Jewish Mode of Slaughtering at the City of Melbourne Abbattoirs [*sic*]."

86 "Mode of Slaughtering Cattle at the Abattoirs," September 14, 1882.

a "glaring case like this of downright cruelty." The writer called upon the SPCA (or the APS more accurately) to take measures against shechita and "to make them introduce a little spicing of humanity into their mode of slaughtering."

The allegation by Humanitas highlights the two principal contentions in such cases. There is an allegation that the Jewish method is cruel per se. In this case Magistrate Buchanan had rejected any attempt by the defense to offer an argument that shechita was less cruel, or no crueler than the method of pithing commonly used in non-Jewish slaughter at the Glebe Island abattoir. For him, the question was simply whether the Jewish method was cruel. By framing the issue in this way, the Sydney magistrate eliminated from legal discussion and argument claims that would arise in other cases comparing the Jewish method with stunning, pithing, the use of bolts etc. Buchanan's position prevented any discussion of the broader moral issue that killing is cruel and that debate about comparatively less cruel methods of slaughter must always be the only relevant consideration in such cases as long as killing animals for human consumption is accepted and legal. The only concession made by the Balmain magistrate was his recognition that animal cruelty statutes did not apply in cases where the animal was killed for a lawful purpose, and the implicit acceptance that human consumption of meat was a lawful purpose.

The judicial framing of the case, with its rejection of any comparative discussion, and the magistrate's personal comments about the cruelty of the process, combined with the vagueness surrounding the idea of a defense based on a practice made lawful by custom, allowed writers such as Humanitas to conclude that Philip Philipstein had avoided conviction not because shechita was not cruel, but because the Jewish practice was somehow exempted from the operation of the law. This meant that Humanitas and others could condemn the legal process for privileging a cruel religious practice, while at the same time boldly asserting that the Jewish method of slaughter was inhumane. The dismissal of the charges in this case confirmed the worst fears of defenders of Jewish religious freedom and the practice of shechita by situating the Jewish method of slaughter as an exception to the general rule prohibiting the infliction of ill-treatment on animals. Judaism and the Jewish practice in this instance were condemned because Jews were refusing to abide by the wishes of the Australian public expressed in "the humane laws of this country."

These remarks were echoed in the pages of editorial writing from the *Sydney Morning Herald* in response to Rev. Davis' letter. The letter was portrayed as a confession that the Jewish method was inhumane. The writer condemned the magistrate's reasoning about the lawful purpose and the practice having been legalized by custom, and blamed Buchanan, who "allowed himself to be influenced in that direction by the Rabbi."[87] The *Sydney Morning Herald*, the paper of John Fairfax, founding member of the APS, urged the strict application of the law, and the rejection of Buchanan's reasoning. Further prosecutions were called for because "it is the duty of the authorities to act on their own responsibility in the matter, and to see that such cruel practices … in this case are not repeated."

Louisa Meredith, former honorary secretary of the Tasmanian SPCA entered the New South Wales fray in a letter to the Fairfax paper condemning shechita in no uncertain terms.[88] That "horrors so gigantic" as those described by the prosecution at the Balmain Water Police Court could continue with the judicial seal of approval, under the guise of religion, was "humiliating and disappointing." The practice of suttee, she reminded readers (invoking a well-worn comparison) was not permitted, despite "the fiercest priestly fanaticism." How could such "sickening, revolting details of the case" not have been treated in the same manner? Cruelty in the name of religion was and would remain cruelty and had to be dealt with under the law. This was a repetition of the arguments put forward by others, and is inherent in the discursive dynamics of anti-shechita cases that present the Jewish method of slaughter within the rhetorical framework of presumed unlawful cruelty. They compel its defenders to engage in a process of justification, the terms of which are set in advance, and in the Balmain case, restricted even more by the exercise of judicial interpretive discretion, and by the traditional confines of criminal prosecutions that are always limited to the facts of the case before the court.

When Ballantine managed to invoke broader social, political, legal, and theological principles in the 1855 London trial, he had been able to overturn these natural limitations of legal process and attack the true nature of anti-shechita cases. This did not happen in the Philip Philipstein prosecution, where the magistrate imposed limits on the defense, and proceeded to condemn the cruelty at the core of the accusations against the shochet. This pernicious effect was exacerbated by his decision to acquit on the grounds of lawful purpose, justified and legalized by custom. Meredith joined in the paper's editorial attacks on the magistrate's decision, but she gave voice to the truly pernicious ideas lurking just below the surface in the minds of evangelically fervent anti-animal cruelty campaigners.

87 "The Letter of the Rev. A.B. Davis," *Sydney Morning Herald*, September 20, 1882.
88 "Cruelty to Animals," *Sydney Morning Herald*, September 21, 1882.

> The Mosaic law is essentially one of vengeance, and where possible, of compensation. All the accumulated millions of the Rothschilds could not alas! undo the barbarities of the past, in which hecatombs of God's good beasts have died in unimaginable protracted torture; but would it not be salutary for the souls of all the bodies which have so long been nourished upon the unlawful flesh of agonized creatures to make, as an infinitesimal atom of atonement, a penitential peace-offering? Money is the sinews of war, and a liberal gift thereof from the community now so deeply indebted to the Society P.C.A. for their exposure of their spiritual peril, might avail much in assisting its efforts for the future. I trust that the many wealthy ones of the "chosen people" will adopt my suggestion...

The tone and content of Meredith's letter embody all that is most obviously problematic about SPCA and related campaigns against shechita. The missive is permeated with the ideas and ideals of Christian redemptory evangelical zeal, seeking atonement and the avoidance of "spiritual peril," in the service of ending the suffering of "God's good beasts." This is the blatant, unremitting, and unabashed public expression of the worst antisemitic discourse in circulation at the time. Jews are vengeful, and money grubbing, and these values are at the core of their religious beliefs and practices. The only thing that will make them understand the cruelty of their slaughtering practices, and their entire religious worldview, is the payment of a collective tribute to the SPCA. The stereotype of the wealthy Jew and the social imaginary of the riches of the Rothschilds are not only invoked directly and unashamedly, but they are deployed to sheet home an idea of collective guilt for the inherent cruelty of Jewish religious practice. "Chosen people" is rendered in scare quotes to ensure that the reader understands which set of religious beliefs, atonement, spiritual peril, the salvation of the soul etc., is superior.

"A Christian Member of a Hebrew Philanthropic Society" penned a missive in reply to the *Sydney Morning Herald* that was only partially philanthropic.[89] The writer was pleased to learn that the Jews of Sydney felt even more disgust at the events revealed in the prosecution of Philip Philipstein than the Christians of the city. The correspondent focused on issues surrounding casting, and urged authorities to establish a separate slaughterhouse for the Jewish method, introducing a chain and tackle system for the killing by the shochet. The writer urged the Jewish community to change the way the animal's

89 "How to Slaughter Cattle," September 27, 1882.

throat was cut by adopting a modern two-edged blade instead of the *hallaf* traditionally used by shochetim. He called for greater vigilance and intervention by the SPCA. He urged the end of the system of slaughterhouse inspectors paid by a levy on each animal killed and the introduction of the system adopted in "the fatherland." In the United Kingdom, local boards of health, full time government agencies, had carriage of the supervision and control of slaughterhouses and the adoption of such a method would greatly improve the state of the abattoir.

The writer urged greater supervision of the Jewish method of slaughter in an abattoir specially equipped for the practice of shechita, with new technological advances for the casting part of the killing process, a suggestion that would be part of debates in the United Kingdom for years to come. He urged the Jewish community to operate a fundamental change to the way in which the animal's throat was cut, treading dangerously close to a line of demarcation in relation to religious injunction and practice. The proposal for greater government supervision along the UK lines was unproblematic in so far as it was made for the general purpose of improving all slaughtering practices at Glebe Island for "both Jews and Christians." But the correspondent urged greater vigilance by the SPCA (the APS) and the pursuit of animal cruelty prosecutions as required in relation to the Jewish method of slaughter. "The evil is not traceable to a disposition to cruelty, but to technical ignorance, and parsimony in not sparing money to provide needful appliances and accommodation."

The Animals Protection Society did not take the dismissal of the case against the shochet lying down, although there is no evidence of a further prosecution. It persisted in its public attacks on the "exceedingly tedious and painful" practice of shechita.[90] Its representatives assured readers of the Fairfax paper that "although our society failed to get a conviction, we have not done with the matter, neither do we mean to leave it alone in the future."

Others supported the call for further action, and urged greater vigilance in relation to conditions at Sydney's abattoir more generally where the "most unnecessary, and wasteful incidents in connection to slaughtering for both Jews and Gentiles are of almost daily occurrence." But they emphasized the "revolting and humiliating" nature of Jewish slaughter.[91] Over the next few years, despite assertions by the APS, there is no indication that shechita remained at the top of the Society's agenda or that of the broader public. There were

90 "Jewish Mode of Slaughtering," *Sydney Morning Herald*, March 9, 1883.
91 "Abattoir Humanity," *Sydney Morning Herald*, March 12, 1883.

periodic calls for closer and better inspection of abattoirs, but these arose out of growing concerns over sanitation and public health in relation to diseased meat entering the food supply.[92]

The campaign against shechita would not end with the prosecution of Philip Philipstein and attempts to vilify the practice in the press following the dismissal of the case against him in early days of the Australian summer of 1882. But the records indicate that the use of a criminal summons under animal cruelty legislation was limited to this single instance.

92 "Inspection of Slaughter Animals," *Australasian* (Melbourne, VIC), May 23, 1908.

CHAPTER 7

Outlawing Shechita in America

The SPCA and Shechita in New York

On January 9, 1866, the President of the SPCA, Henry Bergh, wrote to Henry Westheimer, owner of an important Jewish slaughterhouse, denouncing the practice of shechita in his establishment and threatening prosecution if Westheimer did not make changes to the way animals were killed.

The laws of kashrut had been observed since colonial times in New York. Shechita had been practiced so that observant Jews could consume meat in conformity with their religious obligations. Correspondence and other documents, including bills for services rendered submitted by shochetim, evidence the practice of the Jewish method of slaughter in the city from the eighteenth and into the nineteenth century.[1] When Henry Bergh mounted his attack on shechita, he was targeting a long-established religious tradition, and calling into question, in the name of humanity and a Christian god, the rights of Jewish New Yorkers, in a time of great and divisive political and military conflict, to full equality as Americans.

Bergh's language leaves little room for doubt about the sense of moral outrage animating his intervention, or about the deep sense of Christian mission, grounded in his Unitarian ideals, to deal with issues of cruelty to animals that informed his Society's intervention.[2] The practice at Westheimer's slaughterhouse is described as being of "the most barbarous, revolting, and wicked character."[3] According to the leader of the SPCA, killing without

---

1 Petition of Moses Lazarus, Shochet for increase in salary, July 28, 1771; Issac Levi's application to be shochet, August 20, 1794; Application of Alexander Harris to be shochet at $5 per week, January 27, 1796; Letter of Jacob Abrahams, shochet, November 1804, Jacques Judah Lyons Collection, P-15, American Jewish Historical Society, New York.

2 "Treating Brutes Kindly," *New York Times*, March 31, 1879.

3 The letter is reproduced in Friedman, *Tub Taam*, 15–16.

stunning led to the situation in which the laws of New York were being violated on a daily basis. God's laws were being disrespected during shechita. "If you believe that there is a God in Heaven—tremble! For you may be certain that misfortune will overtake you sooner or later, for these infractions of His natural laws."

This first sustained and public attack on shechita in the United States by the SPCA took on not just a juridical character with Bergh's threats to prosecute if practices were not changed. It embodied the outraged sense of Christian sensibility that had informed much of the anti-cruelty movement from its beginnings in England. The struggle for Bergh was not just one involving a disrespect for the laws of New York, but it was clearly a struggle between an Old Testament morality of "the most barbarous, revolting, and wicked character," and a Christian sensibility that sought to eradicate the cruelty inflicted on God's innocent creatures.

Bergh's moral vision and the crusades that concretized it did not single out only Jews for opprobrium. It would bring him into conflict with influential members of his own class. Unlike his Victorian British counterparts in the RSPCA, who were sensitive for the most part to the interests and activities of the upper class, Bergh felt free to attack animal cruelty among all classes. He did not, for example, spare many of the hunting practices of his peers from attack and condemnation. In New York Bergh had a particular focus. Horses were used, and misused, throughout the city, in freight haulage, and in the transportation of humans by railroads, trolleys, and cabs.[4] Bergh was ever vigilant in exposing and seeking to prosecute those who in his opinion abused the animals for the advantage and comfort of human activity. He was a constant thorn in the side of important New York commercial operations, and waged an ongoing and public battle against the railroads under the ownership of Cornelius Vanderbilt.[5]

Bergh embodied this form of Christian endeavor at the heart of the anti-cruelty movement in the early post-bellum United States, characterized by Diane Beers as "the new humanitarians."[6] Bergh was the son of a wealthy shipbuilder. Using his family connections, he had obtained a diplomatic

4 Anna Sewell's "children's" classic, *Black Beauty*, published in 1877, was intended to draw attention to the plight of abused horses in London. *Black Beauty: His Grooms and Companions*, ed. Kristen Guest (Newcastle: Cambridge Scholars Publishing, 2011).

5 "Bergh vs. Vanderbilt: Alleged Cruelty to Horses on the Harlem Railroad," *New York Times*, August 31, 1871.

6 *For the Prevention of Cruelty*, 39 et seq.

posting in Europe. While travelling around the continent he had a Damascene experience, and dedicated the rest of his life, upon his return to the United States, to the struggle for the protection of animals against the wanton cruelty he believed manifested the worst of human experience. He had a broad, and at the same time, narrow, moral framework, but he was dedicated to the public enforcement of his ethical vision. He campaigned, largely in vain, against vivisection as a particularly cruel example of the mistreatment of animals.[7] His rhetorical flourishes consistently invoked ideals of humanity and called upon those who would mistreat "God's dependent and helpless animals" to reform "in the name of God, and the civilization of our generation."[8]

His moral vision defended dumb beasts at every turn. He was a key actor in the extension of prevention of cruelty legislation to encompass the other weakest of creatures, children.[9] But his Christian charity did not extend to those he held responsible for their own actions. He abhorred those who inflicted cruel treatment on animals, but this extended to the criminal classes more generally. While some sought to embody Christian charity and social reform by focusing on the causes of criminality and on the rehabilitation of criminals, Bergh urged a swift and violent retribution, calling for hanging the worst and for the reintroduction of corporal punishment for others convicted of criminal offenses.[10]

His focus was always and consistently on God's helpless creatures. He sought to create in the United States an organization for the prevention of cruelty to animals based on the model of the English RSPCA.[11] He obtained the incorporation of the American Society for the Prevention of Cruelty to Animals under special statute passed by the legislature of the state of New York on April 10, 1866.[12] Four months before its formal, legal creation, Bergh

---

7 "Henry Bergh on Vivisection," *New York Times*, February 2, 1882; "Mr. Bergh on Vivisection: The Sufferings of Animals and Benefits Derived," January 4, 1881; "Mr. Bergh Against Vivisection," February 20, 1880; "Prevention of Cruelty Against Animals: A Card from Mr. Bergh," February 24, 1867.

8 Letters of December 28, 1869, March 1, 1870, to the Directors of the 7th Avenue Railroad; March 1, 1870, to illegible; June 29, 1870, to the Directors of the National Yards, New York, ASPCA Henry Bergh Correspondence, 1869–1871, Museum of the City of New York.

9 "Caring for Children: The Society for the Prevention of Cruelty," *New York Times*, January 20, 1880.

10 "Reformers Much Amazed: Mr. Henry Bergh Attacks Them at Their Own Meeting," *New York Times*, December 1, 1880; "The Lash for Criminals: Shall the Whipping Post Be Set Up Again…," July 29, 1877.

11 C.C. Buel, "Henry Bergh and His Work," *Scribner's Monthly*, November 1878 to April 1879.

12 An Act to incorporate "The American Society for the Prevention of Cruelty to Animals," April 10, 1866, Laws of 1866, Vol. 1, c. 469.

had written to Westheimer under the aegis of the Society threatening a centuries-old religious practice and obligation, motivated by his individual moral position. For much of its early existence, the Society, which spread across the United States with rapidity, would bear the deep marks of Bergh's personal imprimatur and his unrelenting vitality in the pursuit of his vision of justice and retribution.[13]

The first meeting of the American Society for the Prevention of Cruelty to Animals was not held until May 21, 1866, but its work had already begun with its inspectors, under Bergh's close and interventionist supervision, operating in New York from the beginning of the year.[14] The incorporators of the American SPCA, civic leaders all, held their organizational meeting a mere two weeks after the founding legislation was passed.[15] The first perpetrator was brought before the courts of New York City in the days following the passage of the law and fined ten dollars.[16] At the founding meeting, the organizational structure, its by-laws, and systems of operation of the SPCA were formalized. True to the inspiration for Bergh's crusade, the President and Vice-President of the London Society for the Prevention of Cruelty to Animals were named ex officio members of the American SPCA.

Before the first meeting that had formalized its administrative and enforcement structures, the SPCA, largely in the person of Bergh, had successfully lobbied the state legislature in Albany for changes to the legislation governing and prohibiting animal cruelty.[17] An act to better prevent cruelty to animals was passed in the spring of 1866, shortly after the Society's incorporation.[18] The *New York Times* praised the legislative action and supported calls for the creation of a separate society with independent inspectors reporting acts of cruelty to the police.[19]

---

13 Roswell C. McCrea, *The Humane Movement: A Descriptive Survey* (New York: Columbia University Press, 1910).

14 "American Society for the Prevention of Cruelty to Animals," *New York Times*, May 22, 1866.

15 "American Society for the Prevention of Cruelty to Animals," *New York Times*, April 24, 1866.

16 "Cruelty to Animals–First Case of Punishment under the New Law," *New York Times*, April 13, 1866.

17 "American Society for the Prevention of Cruelty to Animals," *New York Times*, April 24, 1866.

18 N.Y. Rev. Stat. 682.2 (1866); L. 1866, C. 682, amending N.Y. Rev. Stat. 1829: Title 6, 26 (1829); Stephen Iannacone, *Felony Animal Cruelty Laws in New York*, 31 Pace Law Review 748 (2011); David Favre and Vivien Tsang, *The Development of Anti-cruelty Laws During the 1800s*, 1 Detroit College of Law Law Review 1 (1993).

19 "The Cruelty to Animals Bill," *New York Times*, April 12, 1866.

> As organizations that relied on the rule of law and that blended public and private authority, humane organizations are typical of an American approach to statecraft. The expansion of government power under private auspices enabled Gilded Age Americans–animal and child protectionists among them–to perpetuate ... their favorite fiction: that even in the midst of ever-increasing "state interference" they remained a people committed to liberty and voluntarism above all else.[20]

Section 7 of the Act incorporating the Society called upon the police in all parts of the state to render the assistance required by agents of the SPCA, and did not include specific powers, as would later statutes, delegating enforcement to inspectors. Agents of the Society would have common law powers of arrest, which Bergh personally asserted on numerous occasions, but they still required assistance from the public police authorities to carry out their mission. This was a constant source of aggravation for Bergh, who did not cope well with the organizational arrangement or the dividing line between public and private in the field of animal cruelty.[21] Nor did he willingly tolerate what he saw as a judicial reluctance to give full force and effect to the punishments available under the law.[22] Despite these impediments, in the first twenty years of its existence, the New York SPCA prosecuted 12,046 cases of animal cruelty.[23]

At the time of the first attempt by the SPCA to classify the practice of shechita as one of inherent cruelty to animals, Jews already constituted a significant presence in New York's meat markets.[24] The rise of industrial slaughterhouses, in the United States, as in the United Kingdom and Australia, was the source of significant agitation from the SPCA and other groups aimed at instigating humane slaughtering.[25] In New York City, the influx of Jewish immigrants had coincided with the operation of decentralized butchering and widespread occupational resistance against attempts to regulate the practice of animal slaughter in the increasingly dense urban center, a battle that would

20 Pearson, *The Rights of the Defenseless*, 19–20.

21 "Henry Bergh and the Police Commissioners," *New York Times*, August 17, 1871.

22 "Cruelty to Animals: Motion to Punish Mr. Bergh for Contempt of Court Denied," *New York Times*, May 29, 1873.

23 "Henry Bergh's Society: What has been Done to Prevent Cruelty to Animals," *New York Times*, January 7, 1886.

24 "General News," *New York Times*, October 11, 1864.

25 Beers, *For the Prevention of Cruelty*, 36 et seq.

continue for several years.[26] This gave rise to the increased activity of public health inspectors and moves to centralize slaughtering and butchering.[27] Concerns about noise, smells, and general nuisances in populated areas, and public health questions, became important social, political, and legal issues.[28] The SPCA lobbied for changes in slaughtering practices. Like its English counterpart, it joined in efforts for the removal of slaughterhouses from populated areas, and in attempts to eliminate backyard and dispersed slaughterhouses, by establishing a single, central abattoir.[29]

In at least one case during public agitation over the hygiene of slaughterhouses, the owner of a Jewish abattoir sought a license to open another establishment, and was met with public opposition and attacks from his competitors. His offense was that in his current operation he had

> ... so successfully guarded the sanitary condition of the neighborhood that the Inspector of the Board recommended and the Board granted the license to him. His business was so successful that he incurred the displeasure of the whole association, and the citizens joined the butchers in a crusade against him. The Sanitary Committee of the Board visited his building, and reported it in a high sanitary condition.[30]

A campaign against animal cruelty and against slaughterhouses and slaughtering practices in the city led to a specific attack on the Jewish method of slaughter. All of this occurred in an urban spatial and demographic context of a growing Jewish population whose religious needs for kosher meat had to be met. Jewish immigrants populated many of New York's areas of urban density. Overcrowding, and poor hygienic conditions, and the concentration of

26 Jared N. Day, "Butchers, Tanners, and Tallow Chandlers: The Geography of Slaughtering in Early-Nineteenth-Century New York City," in *Meat, Modernity, and the Rise of the Slaughterhouse*, ed. Paula Young Lee (Hanover NH: University Press of New England, 2008); "To Prepare for Cholera: Citizens Protest Against the City Slaughter Houses," *New York Times*, March 29, 1885.

27 "Sanitary Reforms: Proposed Removal of the Abattoirs," *New York Times*, March 4, 1874.

28 Day, "Butchers, Tanners, and Tallow Chandlers"; "Animal Diseases and the Public Health," *New York Times*, April 20, 1884; "Board of Health: The Slaughter-House Nuisance," *New York Herald*, February 9, 1869.

29 "An Evil and a Nuisance," *New York Times*, May 10, 1875.

30 "Cowpens and Slaughter Houses: A Spirited Fight Between Jews and Gentiles," *New York Herald*, March 9, 1869.

eastern European Jews would only grow in subsequent years. Not all Jews in the United States and in New York City in particular had always adhered strictly to the laws of kashrut.[31] The issue of labeling meat as kosher and the availability of meat from animals killed according to Jewish law had for some time been an issue in New York.[32] Recent immigrants were followers of more traditional Orthodox practice, and the demand for meat that could be consumed according to religious tenets increased with the population.[33]

Anti-Shechita in New York: Bergh on the Case

Bergh clearly threatened Westheimer with the full force of the law if he did not cease slaughtering without stunning. In January 1866, changes to the New York statute were still four months away and Bergh would have had to rely on the provisions of the 1829 Act that prohibited beating or torturing an animal "maliciously or cruelly."[34] There can be no doubt that Bergh believed that shechita met those legal criteria, but the precedent of the 1855 London case had set out a principle in English law that killing by the Jewish method would not be considered to be unlawful.[35] While freedom of religion jurisprudence was not in 1866 what it is today, Westheimer in the early days after the Civil War would have relied on First Amendment and other protections for the Jewish faith by arguing that shechita practiced in his slaughterhouse was commanded by religious law.

The case did not advance to the formal stage of adjudication. Westheimer instead approached an authority on shechita, Aaron Zebi Friedman, to defend him against Bergh's charges. Friedman was born in Poland and trained from an early age as a shochet. He moved to Germany and served as shochet and rabbi before coming to the United States in 1848. In New York, he acted as shochet at one of the largest abattoirs and developed a reputation for his religiosity and deep learning. In response to Bergh's threats against Westheimer and against shechita in New York, Friedman penned a response, *Tub Taam*,

31 Eli Lederhendler, *American Jewry: A New History* Cambridge: Cambridge University Press, 2017), 39–52.

32 Samuel Oppenheim, "The Question of Kosher Meat Supply in New York in 1813: With a Sketch of Earlier Conditions," *Publications of the American Jewish Historical Society* 25 (1917).

33 Polland and Soyer, *Emerging Metropolis*, 127–129.

34 N.Y. Rev. Stat. Tit. 6 (1829), § 26.

35 The 1849 UK Act prohibited beating etc. that was done cruelly or the torture of animals. Cruelty to Animals Act, 12 & 13 Vict., c. 92 (1849), § 1.

in Hebrew. The work was translated into English, French, and German and is still a point of reference. According to the biographical sketch of Friedman at the beginning of *Tub Taam*, "Rabbi Friedman's able defense in his treatise 'Tub Taam' convinced the Society of the error of their judgment."[36] Berman in his monumental study of shechita makes a similar claim, asserting "Bergh did not express himself against Shehitah thereafter."[37]

While it is true that Bergh does not appear to have proceeded with his threat against Westheimer and shechita, and this inaction did follow Rabbi Friedman's intervention, the incomplete historical record does not demonstrate that Bergh abandoned his opposition to the Jewish method. The exposition of centuries of Jewish law and wisdom did not deter Bergh, who approached his mission to wipe out the ill-treatment of the dumb beasts of God with an undiminished zeal for many years.

The remaining historical record indicates with little ambiguity that Berman's assertion about Bergh's acceptance of the Jewish method of slaughter is incorrect. Bergh and the SPCA continued to pursue and attack the Jewish method of slaughter throughout the second half of the nineteenth century and into the early days of the twentieth. Sometimes they employed direct anti-cruelty prosecutions or threatened/attempted prosecutions of those involved in shechita. At other times, Bergh and the Society followed a more indirect path by insisting that humane slaughter was limited to stunning before killing, or as technology evolved, could only be accomplished by the use of bolt guns to immediately kill the animal before it was butchered.[38] The attacks against shechita were phrased in the familiar terms of humane methods of stunning before killing, or simply of killing, and of ever advancing technological changes that would make the goals of humane slaughter easily achievable. The Jewish method was not just cruel, it was archaic, and insisting on continuing the practice violated ideals of progress through science and technology, as well as modern (New Testament) morality.

Some newspapers contain reports dated the next year that Bergh and the SPCA had instigated the arrest of a shochet in New York on charges of animal cruelty relating directly to their failure to stun the animals before cutting their

36 "Sketch of Aaron Zebi Friedman," iv; Friedman Family Papers; P-948; Box 1, American Jewish Historical Society, New York and Boston.

37 Jeremiah J. Berman, *Shehitah: A Study in the Cultural and Social Life of the Jewish People* (New York: Bloch, 1941), 408.

38 McCrae, *The Humane Movement*, 83–84.

throats.[39] The *Tri-Weekly Standard* focused on Jewish opposition to the move and repeated their claims that Bergh's attack was one aimed centrally at their freedom of religion. The *Charleston Daily News* bulletin was less than complimentary to the SPCA, voicing what was at the time a typical attitude of those who opposed the anti-cruelty movement. "At the instigation of HENRY BERGH, President of the Society for the Prevention of Cruelty to Animals, a Jew "cutter" has been arrested for cruelty to animals by cutting the throat of a bullock instead of first knocking him on the head. There seems to be no end to the follies of those amiable noodles." There appears to be no record of a trial or a verdict. It is likely that the event in question relates to Bergh's ongoing communications with Jewish officials and slaughterhouse owners about putting an end to shechita, and his threats to end the practice by force of law, rather than to an actual prosecution.

Much of Bergh's remaining correspondence from the relevant period is illegible from damage, but one can see from the better-preserved letters received by the ASPCA, that he did continue his agitation against slaughter without stunning in general, and against shechita in particular. Bergh communicated his continuing concerns to many of the relevant actors in New York. His correspondence file contains responses to his letter about shechita from leading representatives of the slaughterhouse industry and different elements within the Jewish communities of the city. What emerges from these engagements with Bergh on the question of the practice of shechita in the abattoirs is both his ongoing desire to impose stunning before killing as the norm, and a clear and robust defense of the practice by those with whom he corresponded.

Mr. Eisnor, owner of a large slaughterhouse at which the majority of the butchers were Jewish, offered a detailed account to Bergh of the latest scientific studies from Europe underlining the fundamentally humane character of shechita, buttressed with references to Jewish religious teachings about kindness and the abhorrence of cruelty. This meant that he, and other slaughterhouses, were more than willing to assist the SPCA in any attempts to lessen animal suffering, but that imposing stunning on shochetim was unthinkable. In one of the earliest and clearest American invocations of the fundamental importance of religious freedom in relation to the Jewish method

39 "Puritanism Redivivus," *Louisville Daily Courier*, May 7, 1867; "The New York Society," *Tri-Weekly Standard* (Raleigh, NC), May 2, 1867; "News Summary," *Charleston Daily News*, April 24, 1867; "News", *Philadelphia Inquirer*, April 22, 1867

of slaughter, Eisnor concluded by stating that he "would solicit your further consideration of the subject before resisting to my reasons which would compel the adoption of a system which would strike at the Religious Liberties of a portion of the citizens of this Republic so noted for its liberal views in this respect."[40]

The editor of the *Jewish Messenger* reiterated the theological and ethical arguments that cruelty to animals was foreign to Jewish teaching and practice. Echoing Friedman's earlier statements about shechita, he asserted no shochet could engage in any acts such as those described in Bergh's correspondence.

> But the mode of slaughtering animals, according to Jewish law, is of the most merciful character and I am unaware and cannot credit the suspicion that in practice there is any departure from so humane a system.
>
> The Shochet, as the official employed to slay the animal is called, must be a man of sterling religious and moral character, well versed in the Jewish law, especially applicable to his official duties—as to killing and as to examining the animal after death, to ascertain that it is sound, for if the lungs or other parts are diseased, it will be prohibited. This is mercy to the living beings who are permitted by law to partake of animal food. Before he is licensed, this official must be examined by a Board constituted for that purpose, in his knowledge of the laws & regulations appertaining to his office and in sharpening his knifes [*sic*], which must be kept free from the slightest notch or roughness of edge.[41]

Two weeks later the editor wrote to Bergh, informing him that, following his previous correspondence, he had attended several slaughterhouses in order to ascertain whether Bergh's continuing assertions of cruelty had any merit. He assured Bergh that he did so, not just to see for himself, but had he found evidence of such acts as those alleged by the SPCA official, he would have done everything in his power to assist the organization in ending such practices. He concluded, "I detected no cruelty, but the cutter actually exhibited tenderness."[42] The editor followed up on his correspondence with Bergh by publishing an open letter to the Jewish butchers of the city. He insisted that some complaints existed both about the casting process and the lack of skill

40 Letter of January 10, 1867, ASPCA, Bergh Correspondence, Letters to Henry Bergh 1867, Museum of the City of New York.

41 Letter of December 4, 1867, ibid.

42 Letter of December 16, 1867, ibid.

and care of certain shochetim. While rejecting the most outrageous of Bergh's accusations, he urged his co-religionists to respect the laws of kashrut and shechita. These complaints were

> ... enough, however, to cause us to call upon you to exercise your vocation with the utmost care and tenderness, in accordance with Jewish law...
>
> We hope the parties addressed will receive our remarks in the spirit in which they are written, and not cause our religious acts to be considered "cruel" when they are really most merciful, and only become otherwise through want of proper feeling on the part of those whom we deem it our duty thus to admonish.[43]

Bergh entered into a similar exchange with the leader of the city's Reform congregation.[44] Henry Bergh was adamant that the SPCA in New York was not alone and that kindred societies throughout the world, as well as "the truly humane everywhere" considered shechita to be "unnecessarily cruel." The Reform rabbi offered a frontal attack on Bergh's assertions about rampant Jewish cruelty in slaughterhouses and about the inherently inhumane nature of the Jewish method of killing animals for human consumption. He referred Bergh to the 1855 Mansion House decision in London that some twelve years earlier found that the Jewish method was not cruel or unlawful. Jewish leaders attempted to put paid to Bergh's factual assertions about the alleged cruelties being inflicted by shochetim in New York abattoirs and to his moral crusade grounded in a fundamental belief that the truly humane of the world condemned the inherent cruelty of the Jewish method. They invoked scientific studies, basic and foundational principles of Jewish law and ethics, and finally the common-law tradition itself. Bergh, who always acted as though he had God on his side, was unmoved, as he continued to pursue his campaign against Jewish inhumanity. The next year, in 1868, Bergh continued his attacks on the "intense cruelty peculiar to the Jewish mode of slaughtering animals," and engaged again in extensive correspondence with some of the city's rabbis, from whom he sought an admission that the Jewish mode was "particularly cruel."[45]

By 1875, the *New York Times* was reporting that the "largest and best conducted slaughter-houses in New York are carried on by Jews, who, strange

---

43 "To Jewish Butchers," *Jewish Messenger* (New York), December 13, 1867.

44 Letter from S. Adler, December 16, 1867, Museum of the City of New York.

45 "New Yorkisms," *Daily Evening Telegraph* (Philadelphia), August 6, 1868.

to say, not only control the cattle market, but apparently monopolize the killing and dressing of cattle."[46] While raising some concerns about the way in which chickens were prepared for slaughter, the correspondent for the *Times* was quick to assure readers that, "it must be said that no people are more kind to animals than the Jews, and that Mosaic laws lay down the most humane rules, which are imperative in regard to killing animals for food."

In the extensive profile of his work in *Scribner's Monthly* in April 1879, some thirteen years after his letter to Westheimer and Rabbi Friedman's response, the magazine's correspondent quoted Bergh in the following terms:

> "I dread to visit those butcher's," said Mr. Bergh one morning, "and have postponed going till it amounts to criminal neglect. Three-fourths of the butchers of the city are Hebrews. Their religion obliges them to bleed to death the animals they slaughter. So they hook a chain around the hind leg of a bullock, jerk up the struggling beast, and cut his throat. Well, their religion doesn't require them to suspend an animal by its hind leg,—which frequently dislocates the hip and lacerates the flesh. This brutal and shocking torture must be stopped."[47]

Bergh did not keep these views to himself, or limit them to occasions when he gave interviews to journalists. In the same year, Bergh visited the large abattoirs on Thirty-Ninth Street and Sixth Avenue in the company of several other officials from the SPCA and a journalist from the *New York Times*. After witnessing the killing of several hogs and cattle, he addressed one of the butchers to inform him "it had been decided at the recent congress in Paris that a blow on the head previous to cutting the throat is the most humane method of killing an animal."[48] The butcher, Mr. Joseph Stern, "demurred" and informed Bergh that for him any meat obtained in the manner suggested would be unsalable, and would be much more likely to quickly putrefy. Bergh moved on to offer criticisms of the ways in which the cattle were hoisted before being killed and warned that the method he had seen tortured the animal.

Anti-cruelty rhetoric and practice from Bergh deliberately confused the killing of the animal, the question of stunning before killing, and the process of casting. The 1867 newspaper reports, and Bergh's 1866 letter to

46 "A Day with Mr. Bergh: Lessons from a Merciful Man," January 31, 1875.

47 Buel, "Henry Bergh and His Work," 875.

48 "A Day in The Shambles: Mr. Henry Bergh Inspects the Slaughter-Houses," *New York Times*, March 29, 1879; "At the Slaughter-Houses," *Jewish Messenger*, March 28, 1879.

Westheimer, as well as the ongoing campaign for so-called humane slaughter, all indicate a basic opposition to shechita because the animal is not stunned before it is killed by the shochet. Bergh's comments to the correspondent from *Scribner's* refer to a "brutal and shocking torture" associated with the Hebrew butchers. His comments to Stern reflect not just the idea that the most humane method of killing involved stunning, but evidence the international transmission of knowledge and norms in the world of anti-cruelty organizations. Given this continuing attack and campaign against the Jewish mode of slaughter, Bergh does not appear to have been convinced or dissuaded by Rabbi Friedman's intervention, or by the content of subsequent correspondence with leaders of New York's Jewish communities. A more generous reading of Bergh's quoted comments in *Scribner's* might limit the "brutal and shocking torture" to the casting process followed by New York butchers. The *New York Times* article, however, offers a clear indication that Bergh's primary concern was with humane slaughter which for him always meant the international standard of stunning. Casting was, would be, and still is, a focus of contention in making shechita preparations as humane as they can be in conformity with Jewish law and with technological innovation. Jewish officials consistently showed themselves as concerned with these issues relating to casting, and as actively involved in attempts to reduce, according to basic principles of Jewish moral and ethical practice, the suffering inflicted on animals.[49] Casting was still important, but it appears to have been second on the list for Bergh.

Bergh's 1879 comments give rise to another objection that informs the analysis of anti-shechita cases. If his views and the facts he related to the *Scribner's* correspondent, combined with those reported in the *New York Times* on casting, were correct and generalizable to the practice of the Hebrew butchers, the animals being killed were not being slaughtered in accordance with binding principles of religious law and practice. Any damage to the flesh or organs of the animal before it was killed, would make it *treif*, not suitable for consumption by observant Jews.[50] It is not unfair to Bergh to suspect that his objection to the Jewish method of slaughter remained one in principle

49 P.S. Pozzi, W. Geraisy, S. Barakeh, and M. Azaran, "Principles of Jewish and Islamic Slaughter with Respect to OIE (World Organization of Animal Health) Recommendations," *Israel Journal of Veterinary Medicine* 70, no. 3 (2015); Moses Hyman, "The Jewish Method of Slaying Animals: From the Point of View of Humanity," *American Jewish Year Book* 25 (1924).

50 "The Hebrew Butchers: Compelled to Violate the Jewish Law– What a Rabbi Says," *New York Times*, February 22, 1880.

and of general application. He believed that slaughter without pre-stunning was inherently cruel and the SPCA under his reign adopted this position with consistency. This does not mean that he would be accepting of any cruelty inflicted during the preparatory or casting stages of slaughter, but it does indicate that this was not his focus in 1866, any more than it was in 1879, when the *Scribner's* correspondent accompanied him through the streets of New York, and the *New York Times'* correspondent visited the abattoir with Bergh and his SPCA delegation, where he highlighted the daily cruelties inflicted on the city's animals for readers.

The available literature reveals no more concrete attempts by Bergh and the SPCA to prosecute those involved in the Jewish method of slaughter in New York. Debate around shechita concerned the community's regulation of the laws of kashrut and the availability and affordability of kosher meats.[51] Concern focused on the efforts of the inspectors of the Board of Health to ensure that all slaughterhouses complied with regulations governing sanitation and food safety.[52] Nationally, newspapers covered the Jewish method of slaughter to inform their readers, and sometimes to praise the humanity and sanitary value of kashrut with its particular attention to the inspection of the animal for disease or damage once it had been killed.[53] As the number of slaughterhouses grew, efficient regulation of shechita would become increasingly problematic for religious authorities in New York City.[54] Issues in relation to the creation of a single and unified regulatory kashrut body, in New York, as in other cases of larger and equally diverse Jewish communities, continued to trouble shechita for some time.[55] The SPCA under Bergh's successors continued (unsuccessfully) to agitate for legislative reform, the aim of which would have been the banning of the Jewish method of slaughtering animals for food.[56]

51 "Meat Inspection," *The Jewish South* (Richmond VA), September 1, 1893; "Beef That Jews Eat," *Los Angeles Times*, March 9, 1893; "A Matter of Faith and Diet," *New York Times*, February 18, 1889.

52 "Kept Within Bounds: How the Business of Slaughter Houses is Conducted," *New York Times*, May 9, 1892.

53 "The Ancient Mosaic Diet," *The Evening Star* (Washington DC), November, 29, 1897; "Food and the Public Health: The Shoket, and Kosher and Trefa Dishes," *New York Times*, February 23, 1896; "How Jewish Meat Is Killed," *Indianapolis Journal*, February 10, 1892.

54 Arthur A. Goren, *New York Jews and the Quest for Community: The Kehillah Experiment, 1908-1922* (New York: Columbia University Press, 1970), 78–81.

55 "Broad Corporation for New York Jews," *New York Times*, February 13, 1912.

56 "Jewish Slaughtering and Humanity," *American Hebrew and Jewish Messenger*, March 6, 1908; "A New Assault on Shechitah," ibid., June 19, 1908; "Reported Attack on New York Shechitah," ibid.; "Bergh Offers $500 Prize: Will Give That Sum for a Humane Device for Slaughtering Cattle," *New York Times*, March 2, 1908.

There is no evidence that shechita was condemned, persecuted, or prosecuted in New York after the flurry of SPCA Bergh-inspired activity in the 1860s and 1870s. At the turn of the new century, however, there would be some important attempts to classify the Jewish method of slaughter as unlawful cruelty to animals by means of the prosecutions of shochetim.

Rocky Mountain Shechita: The 1903 Denver Case

Writing in 1991, the editors of a history of Jews in the western United States could say with some confidence that it is impossible to "know what were the prevailing religious practices among Jews in the West in the nineteenth no less than in the twentieth century. We do not know to what extent and for how long western Jews in small towns struggled to retain Orthodox dietary practices such as *kashrut* in the home."[57]

Like their earlier counterparts in colonial America, and in the post-War of Independence era, those Jews who joined the miners and explorers of the American West no doubt struggled to maintain a strict adherence to the laws of their faith in dietary matters. Ida Libert Uchill in her groundbreaking study of Colorado Jewry reports that the earliest settlers in the territory often quickly abandoned attempts to keep kosher.[58] Later, as Breck finds, many observant Jews refrained from eating meat until one of their number traveled to a major center like Chicago, and returned with kosher goods fulfilling the requests of his co-religionists.[59] The Chicago stockyards complex came into being in 1866, as a result of the expansion of the city as a railroad center and the development of technology that allowed meat to be kept cold for longer periods.[60] By 1901, the 20,000 Jews of Chicago consumed more kosher meat than any city outside London.[61] By the later part of the nineteenth century in large cities such as Denver, the Jewish population was sufficient to support a local

57 Moses Rischin and John Livingston, *Jews of the American West* (Detroit: Wayne State University Press, 1991), 51.

58 *Pioneers, Peddlers and Tsadikim: The Story of Jews in Colorado* (Boulder: University Press of Colorado, 2000), 75, 225.

59 Allen DuPont Breck, *The Centennial History of the Jews of Colorado, 1859–1959* (Denver: Hirschfeld Press, 1960), 61.

60 Jacques Damade, *Abattoirs de Chicago* (Paris: Éditions La Bibliothèque, 2016).

61 Charles J. Bushnell, "Some Social Aspects of the Chicago Stock Yards," *American Journal of Sociology* 7, no. 2 (1901), 159–160; "Where Cattle Are Killed According to the Rabbinical Law: A Special Department in the Chicago Stockyards Where Choice Animals Are Slain Under Hebrew Inspection and Where Jews Have Full Swing," *Los Angeles Times*, March 9, 1893.

kosher economy. There was a practicing shochet there no later than 1880.[62] Kosher meat was shipped from Denver to Jewish communities in more isolated mining settlements in Colorado. In Denver, by the end of the nineteenth and the turn of the twentieth century, the various synagogues operated their own slaughtering practices and appointed their own shochetim. There was apparently no trained porger in Denver, since only the forequarters were consumed by the Jewish community, with the rest of the carcass being sold to non-Jews.[63]

It was in Denver in 1903, ten years before the Halifax case against Abraham Levitt, that a recorded instance of anti-shechita prosecution/persecution in the early twentieth-century United States occurred. On November 15, 1903, the *Washington Post* and the *New York Times* each carried a wire report dated the previous day indicating that Denver representatives of the State Humane Society had stopped the killing of cattle "according to Hebrew rites," because of its alleged cruelty. A meeting was scheduled for the following week between representatives of the local Jewish communities and the Humane Society. The purpose of the encounter was "to evolve some method of killing which will satisfy the demands of the both Hebrews and the humane society officers."[64]

The New York *Sun* gave slightly more detail identifying Humane Society Agent Tuft as the person who issued warrants against Denver's "rabbis" on charges of animal cruelty. Tuft alleged that in the course of killing cattle by the Jewish method, "a knife is used to gouge out their eyes and otherwise torture them before the death stroke is given. He thinks this violates the laws and will endeavor to stop the practice."[65] The regional press throughout the country covered the story and identified the agent as Oliver E. Tuft.[66] They indicated that no arrests had been made and that the district attorney in Denver was considering the matter.[67]

---

62 Uchill, *Pioneers, Peddlers and Tsadikim*, 205; Breck, *The Centennial History of the Jews of Colorado*, 61, 74.

63 "Dispute Over Kosher Meat," *Denver Post*, November 7, 1903.

64 "Kosher Killing Stopped: Colorado Humane Society Objects to Method of Killing Cattle," *New York Times*, November 15, 1903; "Kosher Killing Stopped: Colorado Humane Society Officers Call It Cruel," *Washington Post*, November 15, 1903.

65 "Warrants for Denver Rabbis: Accused of Cruelty in the Slaughter of Cattle," November 7, 1903.

66 "To Arrest Jewish Rabbis," *Omaha Daily Bee*, November 7, 1903; "Says Rabbis Are Cruel," *Minneapolis Journal*; "Cruelty to Animals: Complaint Made Against Denver Jewish Rabbis," *Salt Lake Herald*; "Charge of Cruelty is Filed Against Jewish Rabbis in Denver," *Topeka State Journal*.

67 "An Officious Officer: Interference with the Jewish Way of Preparing Beef," *Arizona Republican* (Phoenix), November 7, 1903.

Significant opposition to the Denver attack on the Jewish method of slaughter came from the New York Jewish communities. The objections to Tuft's actions took on familiar tones. According to New York Jewish representatives interviewed by the press, either the reports were mistaken, or else the shochetim in Denver were completely incompetent and ignorant of Jewish law. No shochet would inflict the injuries described by Tuft on an animal before it was killed since to do so would render it *treif*, not fit for consumption. As Dr. Joseph Silverman, rabbi of Temple Emanu-El, the oldest Reform synagogue, pointed out, "if such things happened ... they were done by some imposter who posed as a 'shochet,' or authorized slaughterer. They could never have been done by anyone connected with the faith, because the laws on slaughter are clearly defined, partly by Mosaic law, and more explicitly in Rabbinical law."[68]

A third explanation was that the Colorado authorities did not understand shechita and that their actions had "been influenced by a few injudicious persons who do not know what 'kosher' killing is." New York Jewish authorities re-asserted the fundamentally humane nature of shechita with the instantaneous killing of the animal with a single continuous stroke of the *hallaf*. They contrasted it with the barbaric method of multiple blows to the animal's head, meant only to preserve the value of the hide with total disregard for the suffering of the beast.[69]

An explanation beyond total incompetence, ignorance of Jewish law and practice, or simply mistaken reporting sprang to the mind of some in New York's Jewish community. By this time, they had almost forty years' experience with agitation by Bergh and the SPCA to put an end to shechita. That option was "that the charges were made by some one hostile to their faith."[70] Given the longstanding hostility of the SPCA and other anti-cruelty organizations to the Jewish method, the idea of an antisemitic motive lurking behind a decision to bring charges against Denver's "rabbis" is not surprising.

The record on the incident remains incomplete. The author of the sketch in *Tub Taam* writes of the "ignorant opposition" to shechita in Denver. "It was then that 'The American Hebrew' reprinted copious extracts form [*sic*] this work and forwarded it to representatives Jews in Denver. The clever exposition

---

68 "Defend Jewish Slaughterers: Hebrews Here Say the Methods Are Not at All Cruel," *Sun* (New York), November 8, 1903.

69 "'Kosher' Method of Killing: What is Said Here of the Order in Colorado to Stop the Butcher's Work," *New York Times*, November 29, 1903.

70 "Defend Jewish Slaughterers."

of the subject and the invincible arguments the book contained, again carried the day, and Denver, like New York, withdrew its opposition."[71]

The *American Hebrew* did reproduce at length the correspondence between Henry Bergh and Rabbi Aaron Zebi Friedman concerning the earlier proposed ban on shechita in New York. The newspaper's accounts repeated the inaccurate claim that Friedman's intervention brought an end to Bergh's campaign for stunning before killing.[72] The editor of the *American Hebrew* reiterated two important elements to the debates then raging in Denver and throughout the rest of the country as a result of the reported arrest of the "rabbis." "Jewish law and Jewish custom have, at all times, tended towards the merciful treatment of dumb animals. The Scriptures abound in injunctions of kindness towards domestic animals."[73]

This is the well-known Jewish response to allegations of animal cruelty during shechita, reflecting millennia of Jewish deontology and theological principles relating to the prohibition of cruelty. In a particularly American twist to the tale of anti-shechita prosecutions and persecutions, the editor drew a radical distinction between Swiss and German attacks on the Jewish method and the American tradition of agitation embodied by Henry Bergh:

> When the anti-Semites of Saxony or of certain cantons of Switzerland set up such charges and caused the passing of such laws forbidding the Jewish method of slaughter, it was, of course, with no other motive than to vex and oppress the Jew. It was otherwise with that noble soul, the late Henry Bergh, through whose initiative mercy to dumb creatures has become a cardinal virtue in our day.

European anti-shechita agitation was the result of an undisguised antisemitic politics. American anti-shechita activities, according to the editorial writer, were the result of well-intentioned, but mistaken, attempts to end or reduce animal suffering. The general attitude of the *American Hebrew* was scrutinized by its rival the Cincinnati-based *American Israelite*.[74] While expressing admiration for the engagement by the *American Hebrew* against the Denver case, the other important organ of the Jewish press was critical of the absence of

71 "Sketch of Aaron Zebi Friedman," v.

72 "The S.P.C.A. on the Wrong Track," November 13, 1903.

73 "Mercy to Dumb Creatures," *American Hebrew,* November 13, 1903.

74 Yehezkel Wyszkowski, "*The American Hebrew*: An Exercise in Ambivalence," *American Jewish History* 76, no. 3 (1987).

significant protest at this resurgence of antisemitism, particularly when it was taken in the context of ongoing European agitation against Jews and the continuing resurgence of the blood libel throughout this period. "It is interesting to notice that in this present year of grace both charges, the letting of blood of both animals and human beings, are professed with lively activity, to the enjoyment of Greek priests, Russian statesmen, German anti-Semites, French bigots, and American asses."[75]

The narrative offered by the *American Hebrew* was consistent with the story that Rabbi Friedman's intervention brought the light of truth to Henry Bergh and caused him to accept the humane nature of the Jewish mode of slaughtering animals for human consumption. But that story was fictional. Bergh continued to agitate for compulsory stunning of animals before killing, and to campaign, indirectly and directly, against shechita well after the Friedman correspondence. Many of his interventions, as reported in the local and national press, expressed his moral revulsion at the practice of shechita in terms invoking God and humanity in equal measure. This does not irreducibly mean that his primary motivation was antisemitic. But the effects of his desire to protect dumb animals and to introduce compulsory stunning were immediately and inevitably anti-Jewish. When combined with Bergh's consistent invocation of his God as the arbiter of morality, of the holy retribution awaiting those who mistreated the creatures who could not speak for themselves, and his clear refusal to accept any argument that Jewish ethics and law forbade wanton cruelty, it is not entirely clear that antisemitism was limited to European anti-shechita law and politics. Bergh's consistent attitude and actions against the Jewish method of slaughter found their natural and inevitable successor in the Humane Society of Colorado in 1903 and in the Society's attacks on Jewish religious practice in Denver's slaughterhouses.

The real objection in Denver, just as it was and had been elsewhere, was to the killing of cattle without stunning. "The society insists on having the animals first stunned by a blow on the brain. Until the courts should decide the controversy 4,000 Hebrews were deprived of meat."[76] The idea of an insistence on killing with stunning came to the surface in the Denver actions of the Humane Society agent, Oliver Tuft. As the interventions of the New York Jewish communities made perfectly clear, Tuft's descriptions of animal torture before

75 "Killing for Food," *American Israelite*, December 10, 1903.
76 "Kosher Killing Not Humane," *Professional World* (Columbia, MO), December 4, 1903.

the single cutting blow from the shochet's blade had no connection whatsoever with the practice of shechita as commanded by Jewish law.

Berman reports that the shochet Wolf Heller was arrested and charged with animal cruelty. Born in what is variously described as Poland or Russia, Wolf Heller immigrated to the United States in 1892. He is listed in the 1910 Census as having been born, like his parents, in "Russia Yiddish," able to speak English, and employed as a "rabbi."[77] After the events of 1903, Heller continued to work as a shochet, although he suffered from periods of unemployment as a result of recurring bouts of tuberculosis.[78]

Berman summarizes the case and asserts that E.H. Kauver (C.H. (Hillel) Kauvar), a long-time Denver Jewish rabbi and prominent American Zionist, called a meeting of lawyers, physicians, and newspaper reporters where shechita was explained. Kauvar was born in Lithuania, but came to the United States at the age of two. He was ordained at the Jewish Theological Seminary where he earned his doctorate. He is widely credited with bringing Conservative Judaism to Denver, although later in life he became affiliated with a more traditional Orthodox practice and observance. According to Berman, local newspapers heaped scorn on the prosecuting authorities and the charges were withdrawn.[79] Other newspaper reports indicate that the Jewish community worked with the Humane Society concerning certain aspects of shechita as practiced in the Denver stockyards, and developed new techniques that allayed the objections of the animal welfare officers, while adhering to the strictures of Jewish law.[80]

Berman's recounting seems to be close to the truth of the Denver shechita persecution. Despite claims that the Jewish community representatives and the Humane Society had reached an understanding on questions of slaughter techniques acceptable to both sides, the Humane Society was not willing to

---

77 Thirteenth Census of the United States, Denver, Colorado, Precinct 4, Ward 16. In 1930, he is referred to as a "meat inspector." Fifteenth Census of the United States, City and County of Denver, Block 896, Sheet 8B. The 1940 census return gives Heller's occupation as a "Jewish Reverend." Sixteenth Census of the United States, City and County of Denver, Block 55, 56, Sheet 10B.

78 JCRS Patient # 844 Wolf Heller, 1905-1909, Jewish Consumptive Relief Society Records, University of Denver; Adrienne Denoyelles, "'Peculiar Resistance': Tuberculosis, Identity and Conflict among Jewish Physicians in Early-Twentieth-Century America," *American Jewish History* 100, no. 3 (2016).

79 *Shehitah*, 408–409. Michael W. Rubinoff, "Rabbi in a Progressive Era: C. E H. Kauvar of Denver," *Colorado Magazine* 54, no. 3 (1977).

80 "New Mode of Butchering," *Herald Democrat* (Leadville, CO), November 17, 1903.

accept just any assurances from the Jews of Denver. E.K. Whitehead, secretary of the Humane Society, issued a statement that was reflective of the long-held attitude in animal welfare organizations. The position of the Society went beyond the specific allegations against Wolf Heller. The basis of its objections and legal action was the refusal of the Jewish community to permit pre-killing stunning. After proclaiming the Society's support of all religious belief and practice, Whitehead went on to indicate the source of the problem. Jews sought a special status and wished to place themselves above the application of the general laws of the state of Colorado. Jews had come to the United States seeking relief from persecution, but the price for the safety and freedom afforded to them in America had to be obedience to American law. The real problem was the refusal of Denver's Jews to allow stunning of animals because of the strictures of Jewish law. For the Society

> ... it is a poor kind of religion which not only permits but enjoins cruelty to dumb animals, for that is what this amounts to...
>
> It is all a matter of religious prejudice or bigotry, as many would call it, but still to be respected unless wicked and cruel. In this case, it is wicked and cruel and distinctly against the law...
>
> If this sort of thing is permitted because of religious prejudice there is nothing which the plea of religion will not excuse. If there is any sort of abuse of dumb animals covered by the law this is, and you may say that whoever practices this kind of slaughter will be prosecuted, and that summarily.[81]

If Jews were not willing to obey the principles of American anti-cruelty laws, they would need to "stop slaughtering altogether." The position that had long informed anti-shechita persecution and prosecutions by animal welfare groups had rarely been more clearly articulated. Jewish practice and religious belief were wicked and cruel. Arguments about religious freedom could not carry the day when religion became the basis for such depravity and for the abuse of dumb animals. The Society had a duty to protect those dumb beasts and to uphold the principles of American legality against foreign beliefs and practices, however defined. The attack against shechita in Denver in 1903, ten years

81 "Views on Jewish Methods of Killing Beef," *Plaindealer* (Ouray, CO), November 20, 1903; "Secretary Whitehead's Views on the Jewish Method of Killing Beef," *Denver Times*, November 7, 1903.

after the Aberdeen case, and ten years before the last prosecution in Halifax, had little to do with the bizarre accusations against Wolf Heller, whatever the formal legal proceedings might have indicated. The Humane Society made it obvious that the case was grounded in its moral position about animal protection and a definition of cruelty in slaughtering practice that demanded stunning before killing. The logical and inescapable conclusion of this insistence excluded shechita from any ethically or legally operative understanding of humanity in American law.

Following the negative publicity surrounding the charges against Wolf Heller, and the angry and outraged reactions from the Jewish communities' religious leaders, the Humane Society was quick to attempt to step back. It argued that the charges against Heller were accusations against an individual and had nothing whatsoever to do with religion. "Religion should never have been injected in the matter at all."[82] Humane Society Secretary Whitehead wrote to Rabbi Kauvar asking him to put paid to the accusation that religion was a motivating factor in the Humane Society's prosecution of Heller. Whitehead continued:

> Will you kindly give as wide publicity as possible to this denial that anybody connected with this office ever said or thought the cruelty charged against Heller had anything to do with the religious rites of the Jewish church...
>
> Nobody connected with this office has said anything to reflect on the Jewish people or religion either in relation to the Heller case or any other.[83]

This statement in support of its agent's actions reiterated the Humane Society's official position, and is irreconcilable with Whitehead's other public declaration that described the Jewish faith as a "poor kind of religion which not only permits but enjoins cruelty to dumb animals." The invocation of the broader religious meaning and motivation of the case against Wolf Heller did not spring from overzealous and misleading newspaper coverage or some kind of overreaction from leading rabbis in Denver. The targeting of Jewish religious practice was specifically the informing factor behind the Humane Society's

82 "'Kosher' Method Was Not Attacked: Prosecution of Heller Personal Affair, Says Whitehead," *Denver Times*, November 7, 1903.

83 Letter from Whitehead to Kauvar, November 7, 1903, reproduced by Kauvar in "The Shechita Question," *American Hebrew*, November 27, 1903.

own public statements, and more obviously still, of their narrative of Wolf Heller's alleged cruelty in the slaughterhouse advanced by Secretary Whitehead.

Denver's Jewish community reacted quickly and forcefully to the assertions and assumptions behind the Humane Society's attacks. Rabbi Kauvar outlined in some detail both the process of shechita and its theological origins and purposes. For the Denver public, he insisted on the humanity of the Jewish method of slaughter, and, as in other instances of anti-shechita persecutions/prosecutions, he emphasized that the acts alleged by the Humane Society were foreign to the practice of shechita and Jewish law. Either the allegations were manufactured for purposes that went beyond an understandable concern for animal welfare, or Wolf Heller was completely incompetent and ignored the religious and moral obligations incumbent upon him as a shochet.[84] As a point of clarity and correction in relation to the national, regional, and local coverage of the events in Denver, Kauvar highlighted that a rabbi was one learned in theological matters and served as the spiritual leader of the community. Heller was not a rabbi, but a "butcher" (shochet), and the terms should never be confused or used synonymously.[85] At a later public meeting, the positions of Jewish dietary laws more generally, and shechita specifically, were explained with an equal emphasis on the humanity of kashrut and on the health benefits for those who consumed only food prepared in accordance with Jewish law.[86]

The long-serving rabbi of Denver's Reform synagogue, Temple Emanuel, William S. Friedman, echoed Rabbi Kauvar's sentiments and expressed the common feeling in Denver's Jewish community.[87] He did so in terms that resonate with responses, reactions, and arguments throughout the history of anti-shechita prosecutions. He pointed out that Jewish law was clear in prohibiting the acts alleged by the Humane Society representative. "Under this law, no maimed, wounded or lacerated animal is fit for food. If animals are injured they are unclean to an orthodox Jew."

Rabbi Freidman made it clear that Jewish law specifically forbade the actions in question and that any slaughtering that took place under such circumstances would not be consistent with the laws of kashrut. The "guileless ignorance" of Tuft in his assertions about the way in which Heller allegedly

84 "Ridicule Charges: Rabbis Resent Statement of Humane Society Agent that Butchers of the Faith Exercise Cruelty in Slaughtering," *Daily News* (Denver), November 7, 1903.

85 Ibid.

86 "Jewish Health Rules Discussed by Rabbi," *Denver Post*, November 22, 1903.

87 "Forty-year Ministry of Rabbi William Friedman Celebrated in Denver," *Jewish Telegraphic Agency*, February 17, 1930.

practiced the killing of cattle for consumption by Jews was manifest.[88] The Jewish community consistently positioned itself by arguing that it could not believe that a shochet could or would act in such a manner, and that acting in such a way during slaughter would take the killing of the animals in question outside the parameter of Jewish law. Friedman added that no Jew would object to a thorough investigation of the matter. Jews not only abhorred animal cruelty, but required the killing of all animals used for food in accordance with the strict norms of Jewish law.[89]

A week after his first intervention, Rabbi Kauvar announced that he would attend the slaughterhouse to witness for himself the practice of the Denver shochetim. Should the cruelty accusations prove founded, he would be the first to voice his support for the Humane Society's efforts to prosecute the malfeasor. On the other hand, if the butchers adhered to the injunctions of Jewish law, "the Humane Society will have an opportunity to learn something to its advantage."[90]

Kauvar offered a detailed account of events as they had unfolded in Denver to the wider American Jewish public in a letter to the *American Hebrew*.[91] On November 13, following the first set of allegations from agent Tuft, Whitehead, Max Klein, Heller's attorney, and Rabbi Kauvar attended the Colorado Packing Company to witness shechita. Unfortunately for the Denver Jewish community, there was a problem with casting the animal, a process that took more than three minutes. "Some of the agents of the Society were extremely zealous, trying to find all sorts of imaginary cruelties. The upshot of it all was that Mr. Whitehead ordered under threat of arrest the Shochetim to slaughter no more unless the animal be first stunned, and thus deprived of consciousness by the 'merciful blow.'"

At the slaughterhouse, Whitehead echoed Tuft's actions, calling a halt to shechita under the threat of criminal prosecution. By insisting on stunning, which he knew was contrary to Jewish law, Whitehead mounted a direct attack on the Jews of Denver. Even accepting that his actions were motivated solely and strictly by a firm moral belief in his duty to protect dumb animals, the direct effect of his actions was to prohibit the Jewish method

---

88 "Jews Never Tolerate Cruelty," *Denver Times*, November 7, 1903.

89 "Colorado State Capital Affairs," *Colorado Transcript* (Golden), November 19, 1903; "Rabbi Denies Cruelty in Butchering Animals," *Herald Democrat* (Denver), November 8, 1903.

90 "Kosher Meat Controversy: Rabbi Kauvar to Investigate Packing House Conditions," *Denver Post*, November 13, 1903.

91 "The Shechita Question," November 27, 1903.

Jewish method appeared "to be a very cruel one." The object of SPCAs was not to enter into theological debate but to act as agents of humanity. "All systems of religion teach that the law of humanity is a Divine law; here we have only to acknowledge that law and to teach it."[14]

Colam's understanding of both theology and natural law is confused and confusing, but at the core of his argument is a dual belief that God commands humane treatment of animals and that the Jewish method of killing animals for food is contrary to that holy command. Another point made by Colam, consistent with his earlier interventions on the criminal prosecution of those involved in shechita is intriguing. He condemned the London prosecution of 1855, not because he did not believe that shechita was cruel, but because tactically and strategically, criminal prosecution was counterproductive. The London case resulted in creating "a feeling of bitterness in a whole religious body, without convincing them of their error." The Jewish method of slaughter was inherently cruel, but criminal prosecution could never have the proper pedagogic effect of convincing those who engaged in shechita, or followed the laws of kashrut in relation to meat, to change their ways. "We must endeavor to attain our end by convincing our opponents of their error." Jews must be convinced to follow "the spirit of an advanced civilization" by recognizing that "their interpretations and prejudices" must "be reconciled with the Divine law of humanity." Science, evidence, rationality (and not too little Protestant theology), all led to the conclusion that shechita was inhumane. Jews must be convinced of the error of their ways. Prosecution under anti-cruelty laws was not the most appropriate way to proceed, but the idea that the Jewish method of slaughtering animals for food was divinely ordained had to be combatted as incompatible with truly Divine law and its command of humanity.

Colam's views were profoundly reflective of the juridical and theologically infused worldview of SPCA activists, especially in the English-speaking world. The positions abjured the racial and religious mélange of overt Continental antisemitism that played a central role in attacks on shechita in Europe. A subtler, more rational, but divinely inspired, belief that Jewish practice was inherently cruel, informed by antisemitic motifs, was at the core of English and American anti-shechita activism and was always at work in attacks on the Jewish method. The ultimate proof of the inhumanity of the Jewish method

14 "Slaughtering with Least Pain to Animals: Debate in the Last Congress of Nations," March 1871, 80.

of slaughter. Rabbi Kauvar insisted that Whitehead was not only attacking Jewish religious practice, but that he was doing so without any practical or real evidence of cruelty. The difficulty with the incident observed on November 13 was one of casting. The process of killing the animal with the shochet's blade had been swift and according to all observers without pain or cruelty.

The next day, on November 14, 1903, another meeting took place between Kauvar and Whitehead in which matters were further clarified. The Humane Society representative declared disingenuously that their objection was not to the killing of the animal by the shochet in accordance with the laws of shechita, but instead with the preparatory process and the suffering inflicted on the animal. The rabbi once again made a clear and unambiguous statement that the process of casting and the particular method used were not directly related to shechita.

> It goes without saying that, according to the Jewish law, the method of preparing is to be as humane as it can possibly be made.
>
> A further demonstration of the Jewish method of slaughter was arranged at another slaughterhouse on November 16. Four animals were killed, and the casting lasted no more than forty-five seconds in the longest case.
>
> Having demonstrated that even the method of preparing the cattle for the knife there need be no unnecessary pain to the animal, I had also behest that the Society will not in the future attempt to interfere with the Shechita. The Humane Society of the State of Colorado has convinced itself that the Jewish method is ideal…

While reports of Tuft's original complaint against Heller and shechita come from late October and early November, official court documents record a more detailed formal accusation lodged by Tuft on November 25, 1903. He swore a complaint affidavit against Wolf Heller for alleged acts of cruelty occurring at the slaughterhouse on October 5.[92] The official legal documents alleging that Heller had engaged in acts of cruelty came more than one week after the series of meetings and demonstrations of casting and shechita attended by Rabbi Kauvar and representatives of the Humane Society, at the end of which Kauvar claimed that the Society was now convinced of the humane nature of the Jewish method. The time lag between the alleged acts of cruelty and the

92 The People of the State of Colorado v. Wolf Heller, No. 34036-1, County Court, Denver.

instigation of these official legal proceedings serve as some indication that this was more than the spontaneous decision of a singular enforcement agent. The intervention of the Humane Society Secretary with his broad claims against the Jewish method's inherent cruelty and inhumanity support the view that the Denver case amounted to an unambiguous attack by animal welfare officials against shechita.[93] Whitehead and the Society, despite the demonstrations offered by Rabbi Kauvar and the shochetim of Denver, persisted in pursuing Tuft's claims against Heller. The facts alleged by Tuft, and affirmed by the local District Attorney, related to the acts preparatory to the killing of the cattle in question. This was another prosecution of a shochet in relation to the process of casting, and not one directly attacking the act of killing without prior stunning. Throughout the process of negotiation and demonstrations, Whitehead had insisted that the failures of casting had been the most troubling factor for the Humane Society, despite his broader claims against the Jewish method and the consistent attitude of organizations such as his towards compulsory stunning.

Heller was alleged to have kept the animals in close proximity to other animals as they were being killed. It has long been the practice proposed for humane slaughter that animals not witness the slaying of their fellows because of the distress that this causes. Further allegations were made against Heller, asserting that in the casting process animals were choked by ropes around their necks; in moving the animals into the upside down position to allow their throats to be slit and the blood to leave their bodies, the apparatus consisting of a steam derrick, a chain, and hooks, Heller ill-treated the animals causing them to thrash about; that during the thrashing, they smashed their head and horns on the floor causing them to bleed profusely; Heller himself, and others on his orders, gouged out the eyes of the animals as they lay suspended. All of this was reflected in the more detailed newspaper accounts of the case, and caused respected rabbis in New York to assert that if true, the process described was in clear violation of the laws of kashrut and the damage caused to the animals, "bruised and cut and gashed" would render them unfit for consumption under Jewish law.

The Humane Society appears, despite its negotiations with Rabbi Kauvar and its attendance at demonstrations of the Jewish method, to have taken

93 The sub-heading of one newspaper account clearly reflects this understanding. "Kosher Meat Controversy... Secretary Whitehead's Reiterations of Agent Tuft's Report on Slaughtering Methods Stirs Up Jewish Population," *Denver Post*, November 13, 1903.

the position that its allegations against Wolf Heller were true, that Tusk's description of events in the slaughterhouse were accurate, and that whatever position it might wish to adopt on the Jewish method generally, it would proceed with its case against the shochet.

On the evidence of Tuft's affidavit, the District Attorney charged Heller with having unlawfully tortured, tormented, cruelly beaten, and needlessly mutilated the cattle. The shechita case in Denver in its official judicial documentation embodies the melding of the public and the private that characterized the entire history of such prosecutions. The sole evidence against Heller at this stage was the affidavit of Tuft, an agent of the Humane Society. The official charging document came from a public official, the District Attorney, and embodied the rhetorical frame of animal protection legislation, indicating that the alleged acts of animal cruelty were an attack on the "peace and dignity of the people of the state of Colorado." Heller was formally arrested and placed on a three-hundred-dollar bond on December 14, 1903, when Rabbi Kauvar and Whitehead were thrashing out a supposed compromise. Further court documents indicate that witnesses were summoned to trial in May and June 1904 and one received a payment of forty dollars and fifty cents from the state for seven days' attendance, but the official court records contain no evidence of a trial, and no report of a conviction or acquittal, nor is there is there any newspaper report of such a case. The official paper trail ends there.

Animal protection societies continued to campaign throughout the United States for better conditions in slaughterhouses and for more humane methods of slaughter.[94] "A woman preacher" had been active against the Jewish method in Detroit at the same time as the case was brought against Wolf Heller in Denver.[95] As a result the Health Officer in Detroit indicated that he intended to take action. The Detroit Jewish community formed a special committee who met with the Health Officer and explained to him the practice and theological origins of the Jewish method. After the meeting, the officer ordered a copy of Dembo's scientific study of shechita. The following day he and the Meat Inspector visited the slaughterhouse with members of the Jewish community and witnessed the Jewish method. William Saulson, head of the *kehilla*, reported that after the visit by Detroit officials to the slaughterhouse, the Jewish community "heard no more about it" and the matter was ended.[96]

---

94 "Jewish Slaughtering and Humanity," *American Hebrew and Jewish Messenger*, March 6, 1908.
95 "The S.P.C.A. on the Wrong Track," *American Hebrew*, November 13, 1903.
96 "Shechitah in Detroit," *American Hebrew*, November 27, 1903.

Anti-cruelty campaigners did not completely abandon the idea of ending the Jewish method of slaughter. In Pennsylvania, the SPCA investigated conditions in abattoirs and focused its attention on claims that shechita was particularly inhumane.[97] SPCAs lobbied state legislatures for measures imposing the compulsory use of stunning. Ohio SPCAs in both Cleveland and Cincinnati spearheaded attempts to both persuade Jewish butchers to abandon shechita and to convince legislators to outlaw the practice, again without success.[98]

At the same time, the SPCAs and Humane Societies continued to deplore the overall conditions and practices in American slaughterhouses, alongside their continuing attacks on killing animals for human consumption without stunning. Their key opponents on the former front were "the packing interests" and those on the latter "the defenders of the Hebrew method." While continuing to deny any antisemitic motivations, proponents of a ban on shechita continued to insist on the basic cruelty inflicted in the Jewish method of slaughter.[99]

Atlantic City and New Jersey Shechita: A Forgotten Case?

Berman in his study of the Jewish method of slaughter in America records failed attempts in Massachusetts in 1911 to introduce compulsory stunning before killing in the commonwealth.[100] A later attempt to introduce a generalized prohibition of killing in the absence of stunning took place in Connecticut in 1929, but was withdrawn.[101] For Berman, the Denver case stands out as the closest the courts came to criminalizing shechita by applying general animal cruelty legislation against shochetim in the United States. The prosecution of Abraham Levitt in Halifax highlights the ways in which the Jewish community was able to counter the attack on shechita and to overturn the conviction of the shochet.[102] But Berman's history of "anti-Shehitah incidents in America" is incomplete. This chapter concludes with a case that went unnoticed by

97 "The S.P.C.A. and Cattle," *Philadelphia Inquirer*, November 1, 1905.

98 "To Oppose Kosher Butchering," *American Israelite*, August 22, 1912; "Cleveland S.P.C.A. Starts Attack on Shechita," *Jewish Advocate* (Boston), May 10, 1912.

99 Francis H. Rowley, *Slaughter-House Reform in the United States and the Opposing Forces* (Boston: Massachusetts Society for the Prevention of Cruelty to Animals, c. 1913), 9 et seq.; "Humane Slaughter," *Boston Sunday Post*, June 30, 1912.

100 *Shehitah*, 409, 411.

101 "The Shechita Seems to Stand the Test of Time," *Jewish Advocate*, May 16, 1929.

102 *Shehitah*, 409–411.

Berman and seems to have passed into the sphere of historical amnesia in relation to the story of attempts by humane societies to attack shechita by criminalizing it under general anti-animal cruelty legislation. But in March 1910, a new case emerged. "On the complaint of the local Society for the Prevention of Cruelty to Animals, a magistrate of Atlantic City, N.J., fined Morris Kaplovitz for slaughtering sheep by the Jewish method."[103]

The New Jersey SPCA is the second oldest in the nation, founded in 1868 by act of the state Legislature.[104] By 1910, there was a well-established Jewish community in southern New Jersey and in particular in Atlantic City. By the second decade of the twentieth century, nearby Philadelphia had a Jewish population of some 200,000, and Atlantic City itself had 4,000 Jews among its 50,000 citizens.[105] In addition to the urban population in Atlantic City proper, many Jews established agricultural enterprises in rural Atlantic County and Jewish farming colonies were found neighboring Cumberland County.[106]

The first synagogue, Beth Israel (Reform) was founded in 1893. The Orthodox congregation in Atlantic City, with which Morris Kaplowitz (Kaplovitz) was associated, Rodef Shalom, came into existence in 1896.[107] Kaplowitz served as cantor *(hazan)*, and it would appear, as shochet. His occupation was listed in the 1910 federal census as "prayer synagogue" and he served in this capacity for many years before his death in 1939.[108] In the week in which Morris Kaplowitz went to jail for animal cruelty, Rodef Shalom celebrated the dedication of its new synagogue.[109]

Many of the details that might have been available from original sources are now missing, including the original court documents.[110] The Kaplowitz

---

103 Albert M. Friedenberg, "The Year 5669," *American Jewish Year Book*, 11 (1909–1910): 62.

104 An act to incorporate the New Jersey Society for the Prevention of Cruelty to Animals, L. 1868, c. 335; An act for the prevention of cruelty to animals, N.J. Rev. Stat. 64-82 (1873).

105 Samson D. Oppenheim, "The Jewish Population of the United States," *American Jewish Yearbook* 20 (1918–1919): Table A, 49 and Table C, 52.

106 Deborah E. Popper, "'Great Opportunities for Many of Small Means': New Jersey's Agricultural Colonies," *Geographical Review* 96, no. 1 (2006); Ellen Eisenberg, *Jewish Agricultural Colonies in New Jersey, 1882–1920* (Syracuse: Syracuse University Press, 1995).

107 Leonard F. Vernon and Allen Meyers, *Jewish South Jersey* (Charleston, SC: Arcadia, 2007), 7.

108 Thirteenth Census of United States 1910, Atlantic City, Ward 4; Obituary, "Morris Kaplowitz," *New York Times*, March 22, 1939; "Kaplowitz, Morris," Jacob Rader Marcus and Judith M. Daniels, *The Concise Dictionary of American Jewish Biography* (Brooklyn: Carlson Publishers, 1994), 314.

109 "Tenth Year of Rodef Sholam [*sic*]," *Atlantic City Sunday Gazette*, March 14, 1909.

110 Gina Y. Holmes, Court Administrator, Atlantic City Municipal Court, personal correspondence, May 10, 2017.

case was significant enough to have merited mention in the *American Jewish Year Book*, and it did receive attention at the time in newspapers, both in New Jersey and beyond. That newspaper coverage reveals some of the pitfalls and difficulties of relying on such accounts for an accurate legal historical telling of the story.

The Atlantic City shechita case went out on the newspaper wires and featured in publications as far afield as Indiana, Kansas, Mississippi, and Missouri.[111] The headlines and stories in these out of state papers indicate that for the public, the case was one that involved the killing of an animal according to the Jewish method of slaughter. Rabbi Kaplowitz was accused of animal cruelty as a direct result of performing shechita, bringing the case into the long history of attempts to use general anti-cruelty statutes to prosecute shochetim carrying out a practice commanded by religious obligation. But the newspaper coverage is replete with contradictions and uncertainties and caution is required. The stories from Indiana and Mississippi state that Rabbi Kaplowitz killed a calf brought from a local farm. The accounts from Kansas and Missouri identify the animal as a sheep. There are claims that he was charged with the cruel killing of the animal in each account. The Missouri story identifies the place of the killing of the animal not as Atlantic City, but as Philadelphia.

Local New Jersey newspapers covered the story and committed some of the same mistakes as their farther flung colleagues. The *Ashbury Park Press* claimed that "the calf was brought here from the mainland by a farmer of the Jewish faith to be killed according to the prescribed rites of the Hebrew religion."[112]

The *Trenton Evening Times* reported that Rabbi Kaplowitz was convicted of cruelly killing a sheep.[113] The local newspapers in Atlantic City, the *Daily Press* and the *Evening Union*, may have been the sources of the confusion

---

111 "Kills Calf in a Religious Ceremony and Atlantic City Women Kick: It Means a Court Fight," *Palladium-Item* (Richmond, IN), March 10, 1909; "Rabbi Spends Hours in Jail: Refuses to Pay Fine on Conviction of Cruelty to Animals," *The Hutchinson News* (KS), March 10, 1909; "Rabbi Accused of Cruelty: Faces Trial for Killing Calf According to Rules of Church," *Jackson Daily News* (MI), March 9, 1909; "A Rabbi Goes to Jail: For Having Allowed a Sheep to Die on a Philadelphia Lot," *The Sedalia Democrat* (MO), March 10, 1909.

112 "Rabbi Accused of Cruelty: Atlantic City Woman Causes Arrest for Killing a Calf," March 8, 1909.

113 "Rabbi Jailed for Killing of Sheep: He Chooses Cell Rather Than Fine, Denying He Was Cruel," March 9, 1909.

found in more distant accounts. The *Daily Press* referred to the complainant as "Mrs. Waner." It recounted that the cruelty occurred when Rabbi Kaplowitz killed a calf belonging to a farmer named Simons who had brought the animal to Atlantic City "to have it killed in accordance with the doctrines of his faith."[114] The *Evening Union* on the other hand produced a lengthy account of the evidence in the case, including "Mrs. Warner's" specific allegations of cruelty in the killing of a sheep in a vacant lot in Atlantic City.[115]

Whether Rabbi Kaplowitz killed a sheep or a calf, it is accepted in all accounts that a fundamental aspect of the charges brought against him is that the "woman agent" of the SPCA who brought the case to the magistrate did so because she claimed to have witnessed the unnecessary suffering of the animal as a direct result of shechita. Some stories offer more detail about what the female agent, "Mrs. Waner," "Mrs. Nellie Warner," or "Mrs. Anne Warner" saw. Others assert that while Mrs. Warner claimed to be a volunteer agent of the SPCA, the organization itself was reluctant to become involved in the prosecution.[116] Her allegations included statements that the animal took between thirty and sixty minutes to die, and that when she intervened, the Rabbi's assistant tore out the animal's windpipe and struck her with it. Given the differences among the accounts of the case, and the absence of judicial records, it is perhaps worth focusing on elements that appear most credible and fit with the overarching themes and narratives that have emerged from other anti-shechita cases.

One common aspect is that the accusers claim to have witnessed extended periods of thrashing and suffering by the animal after the shochet delivered the fatal cut. Three explanations emerge from such accounts and from the testimony of Mrs. Warner in the Atlantic City case. There is the possibility that concerns over the welfare of the animal might lead to an exaggerated account of the so-called death throes of the beast. Evidence from the newspaper stories in this case give the time frame as somewhere between a half hour and an hour for the calf (or sheep) to expire. Such a discrepancy on a key element of the cruelty charge might raise significant issues as to the accuracy and credibility of the prosecution witnesses. A second explanation for the evidence in relation to prolonged thrashing by the animal after its throat had been cut is that offered by many scientific experts who testified in other anti-shechita cases that what the witness saw and believed to be indications of suffering by a still

114 "Rabbi's Arrest Stirs Trouble," March 8, 1909.

115 "Rabbi Jailed in Default of Fine," March 8, 1909.

116 "Rabbi Accused of Cruelty," *Jackson Daily News*.

living and conscious animal was a series of muscle movements, independent of consciousness or pain. The third explanation in this case is that what the witnesses recounted, even taking into account the time discrepancy, was an accurate representation of what happened. In such a case, the shochet would have failed to perform his duties in carrying out shechita properly, and all those concerned with animal suffering would applaud the prosecution.

What does emerge from the available facts of this case and the accounts of the evidence presented to the magistrate, is that Rabbi Kaplowitz was an experienced and skilled shochet who consistently denied that he had treated the animal cruelly. For the shochet and for the Atlantic City Jewish community, the case brought by Mrs. Warner was little more than an ill-disguised attack on shechita. "Local leaders of the faith are excited over the arrest, and expert lawyers have been retained to support the contention of the rabbi that the killing of the calf was done in a manner allowed by law in deference to the religious rules of his sect."[117]

So adamant that he had acted according to the tenets of his faith and the laws of kashrut was Rabbi Kaplowitz that, upon his conviction and the imposition of a ten dollar fine, he refused to pay, and while insisting that he would appeal to the highest courts, he elected to spend several hours in jail. "He claims he has broken no law, and will secure a decision giving him warrant to kill animals according to the rules prescribed by the Bible."[118]

Rabbi Kaplowitz was represented by prominent local attorney, Henry Schneider, who was New Jersey Commissioner for Inland Waterways. After a failed attempt to obtain a writ of habeas corpus because the local judge was ill and unable to hear the case right away, Schneider paid the fine under protest to free his client from jail, promising to appeal.[119] Sadly, the trail of the Atlantic City shechita case ends here. There is no further evidence that it proceeded to a higher jurisdiction, but equally there is no indication that any further action was taken to criminalize the Jewish method of slaughter in New Jersey during the relevant period. Magistrate Hughes obviously found that there was sufficient evidence to convict Rabbi Kaplowitz of cruelty in this particular instance, where he appears to have found the evidence of animal suffering convincing. Just as in the conviction of Abraham Levitt in Halifax, the magistrate in the absence of scientific testimony, took the witness evidence at face value.

---

117 "Rabbi Accused of Cruelty," *Asbury Park Press*.
118 "Rabbi Jailed for Killing of Sheep," *Trenton Evening Times*.
119 "Rabbi Jailed in Default of Fine," *Atlantic City Evening Union*.

He accepted the veracity of the claim reported in the local paper that "the sheep lived for more than a half hour before death finally came to its relief."[120]

Like his Halifax counterpart, Magistrate Hughes gave little credence to the rabbi's consistent claims that he acted according to the laws of his faith and that those laws prohibited the infliction of cruelty on animals during shechita. The worst interpretation is that for the first time in North American legal history a shochet was convicted of animal cruelty in carrying out the slaughter of an animal for food under the laws of kashrut. A slightly better interpretation is that what Rabbi Kaplowitz did in Atlantic City in March 1909 violated the rules governing the practice of shechita. He failed to properly kill the animal with the single blow as required, and as a result did treat the animal cruelly. If the animal lived for a half hour after the cut, as witnesses claimed, this would be a clear indication that Rabbi Kaplowitz had failed miserably in his religious duty to kill it according to the precepts of Jewish law. Such an understanding would place the conviction of the rabbi outside the strict limits of the overt criminalization of a religious practice. But to arrive at this conclusion, it is necessary to ignore Rabbi Kaplowitz's consistent protests that he acted according to the laws of his faith, his lawyer's adamant argument to this effect, and Kaplowitz's insistence on going to jail rather than pay a fine, as well as what science tells us about involuntary muscle movement after death. These elements are consistent with an understanding of the case as a direct attack by the SPCA or its agent on the practice of shechita.

In his recent study of the Atlantic City communities, Gordon highlights the evolving nature of the Jewish presence, and in particular a continuing degree of animosity between the early "German" Jews, and the more recently arrived "Russian" Jews associated with Rodef Shalom.[121] Reports of this case indicate that local community leaders were troubled and concerned over what they experienced as a direct attack on their religious practices. But they could not present a united front in their response. A leader of the German congregation wrote to the editor of the *Daily Press* to express the concern that the newspaper had committed a basic error in its coverage, one that posed a great risk to Jews in Atlantic City. The concern was not about reports of cruelty to animals, or about allegations that shechita might be inhumane. Instead it was about the use of the honorific– rabbi. "Permit me to state for your benefit and that of your readers that Mr. Kaplovitz is not and never was a Rabbi nor did he

120 "Rabbi Jailed in Default of Fine."

121 Jerry Gordon, *From Its Beginnings: The Jewish Community of the Atlantic City Area* (Atlantic City: published by author, 2016), 5–6.

ever lay claim to such title."[122] The next day, the President of Rodef Sholem replied by informing the readers of the newspaper, and Rabbi Fisher, who was obviously unaware of developments in the rival synagogue that "Mr. Kaplovitz is a Rabbi. The congregation accepted him as a rabbi three years ago."[123]

This disagreement, at the margins, but perhaps closer to the heart of the Atlantic City shechita case, does carry important echoes of other instances and debates within Jewish communities subjected to attack in relation to their religious observance. Some dismissed Abraham Levitt in Halifax as "not a rabbi," or as a mere veterinary cutter. Wolf Heller's job as a shochet in Denver was clearly distinguished by Kauvar from the status of a rabbi. This can be understood not just as a hierarchical assertion of rank, but as an attempt to place the shochet, Levitt, Heller, or Kaplowitz, at the margins of Jewish community leadership, where technical incompetence might come into play. There might be genuine concerns about actual skill and experience, about shochetim who were competent with smaller animals such as chicken, but not as skilled with larger sheep or cattle. This would mean that debate was shifted, as in the Myers and Friedner cases in England, to focus on issues of skill, and to draw attention away from the theological core of shechita, especially in the minds of a general public, or even a governmental and legal elite, not to mention ardent SPCA members and inspectors.

Rabbi Kaplowitz insisted throughout these few days that his prosecution was unfounded and that he had done nothing but follow the laws of his faith. The Atlantic City Jewish community was obviously concerned with the potential implications of a broader attack on shechita. The case was significant enough to have warranted a mention in the *American Jewish Year Book*. The absence of follow up by local synagogue officials, or of a broader campaign that accompanied other instances of anti-shechita prosecutions, gives some support to the ideas that this was not really a shechita case. More speculatively, this was a case of individual initiative by Mrs. Warner and not an organizational effort by the SPCA. Perhaps there was a political arrangement to permit the Jewish method of slaughtering animals for food to continue. It seems we may never know. What we do know is that for all its ambiguities, contradictions, and lack of clarity, the prosecution of Morris Kaplowitz in Atlantic City in March 1909 appears to be the first recorded instance of the conviction of a shochet for animal cruelty in North America.

---

122 "Kaplovitz Not Rabbi," Letter from Henry M. Fisher, March 9, 1909.
123 "Defendant A Rabbi, He Says," Letter from A. Phillips, March 10, 1909.

CHAPTER 8

The Massachusetts Anti-Shechita Story

The Massachusetts SPCA and Shechita

During the prosecution of Abraham Levitt in Halifax, Nova Scotia, in the spring of 1913, Magistrate Fielding adjourned the case to allow the defense to produce religious authority for the practice of shechita and to permit prosecutor Murray to present what he claimed were two older Massachusetts cases of a similar nature.[1] The connections with the "Boston states" were strong at this point in Nova Scotia history. As the facts of the case against Rabbi Levitt and others make clear, anti-cruelty societies participated in knowledge exchange networks and were kept informed of developments in other jurisdictions. The Massachusetts SPCA was particularly active at the time and powerful in the struggle to promote and ensure animal welfare, and to end killing without stunning.

Founded in 1868, the Massachusetts SPCA was the second most influential in the nation after Bergh's New York group. While inspired by events in New York, the leader of the Massachusetts Society, George T. Angell, acted autonomously and would for many years be a leading figure in anti-cruelty circles and in the humane education movement.[2] The son of a Baptist minister, Angell became a successful attorney in Boston before turning fulltime to his humane work. As founder and President of the Massachusetts SPCA, the American Humane Education Society, and the Parent American Band of Mercy, Angell wrote and travelled extensively throughout the United States, delivering important speeches on a vast array of topics. He travelled to Europe, where he attended international congresses of the anti-cruelty movement, and befriended leading animal welfare activists in Switzerland, Germany, Italy, Austria, and the United Kingdom. In England, he became close to John

1 "Jewish Rabbi Fined by the Stipendiary," *Halifax Herald*, April 22, 1913.

2 George T. Angell, *Autobiographical Sketches and Personal Recollections* (Boston, 1893).

Colam of the RSPCA, with whom he maintained a regular correspondence throughout his career.

He created the Massachusetts SPCA's publication, *Our Dumb Animals*, the first of its kind, and saw to its mass circulation throughout the United States and beyond. The journal provided a powerful forum for his animal welfare work, and served to reinforce international contacts and networks of knowledge. It regularly contained information about international animal welfare work.[3] His obituary in the *New York Times* noted his activism and untiring efforts in the field of animal welfare. "Mr. Angell had been actively engaged in humane work since. In one year, he had published more than seventeen millions of pages of humane literature. He traveled throughout the United States and Europe in pursuance of his work. On these travels, he caused to be established more than seventy thousand 'bands of mercy.'"[4]

Among the subjects that greatly concerned Angell and the Massachusetts SPCA were the conditions of the transport and slaughter of animals for food, and related questions surrounding hygiene in abattoirs. *Our Dumb Animals* featured reports from the United Kingdom in particular on the latest developments in animal welfare in slaughterhouses, along with studies of animal transport for slaughter and conditions in American abattoirs.[5] The question of shechita was not long ignored by Angell and the Massachusetts SPCA, but the Boston lawyer abjured, at least formally, the direct and unrelenting attacks on the Jewish method that characterized European animal welfare agitation.

The first intervention on the shechita question in Massachusetts was a translated report from the international convention of anti-cruelty organizations in Zurich in 1870. The body of the article relates to the discussions and controversies that arose at the Swiss meeting between those seeking to characterize the Jewish method as inherently cruel and religious authorities who defended shechita on both theological and scientific grounds. Bringing the question into the local context, Angell concluded that "the method practiced by the Jews in this vicinity has been brought to our attention, and we respectfully ask some one in authority to give us a description of their peculiar custom."[6]

3 Ibid., 13. Sydney H. Coleman, *Humane Society Leaders in America* (Albany, NY: American Humane Association, 1924), 89–114.

4 "George T. Angell Dead: Known Here and Abroad as 'the Friend of Animals,'" March 17, 1909.

5 "The Slaughter-House," September 1871, 130; "Cattle Transportation in the United States," January 1872, Supplement; "Less Cruelty in Slaughter-Houses," ibid., 168.

6 "Jewish Method of Slaughtering," *Our Dumb Animals*, November 1870, 43.

The Massachusetts SPCA and Angell as its leader sought specific information on the practice of shechita in Boston, without any a priori condemnation. Angell even at this early stage still characterized Jewish practice as a "peculiar custom." This ambiguity in the SPCA's position on the Jewish method of slaughter would remain for some time. Rabbi Dr. Aaron Guinzburg gave a response to Angell's request for an authoritative Jewish intervention. A native of Prague, Guinzburg had taught theology at Newton University and filled rabbinical posts in Maryland, Rochester, and Boston, where he was President of the Warren Street synagogue, Temple Ohabei Shalom, the oldest synagogue in the city, at the time of his death in 1873. He was a renowned scholar and an almost inveterate commentator in newspapers on matters of Jewish interest and welfare.[7] In his first response, Guinzburg offered a detailed examination of Biblical text establishing "that every intelligent reader is able to judge from the few quoted passages, that Judaism is against any cruel treatment of animals."[8] He promised to address the question of shechita specifically in the next issue of the magazine. His declaration was met with editorial ambiguity from the SPCA, a less than objective starting point for animal welfare activists on the Jewish method of slaughter. "The subject has been much discussed in Europe and in this country, and we shall be glad if Dr. G. can relieve the public of the impression that the practice of his people is a cruel one."[9]

The discussion in Europe had been, and would be, led by overtly antisemitic forces, particularly in Germany and Switzerland. In the United States, the issue of shechita in 1870 was at the forefront of Henry Bergh's New York campaign for humane slaughtering which by definition included stunning. The discussion was shifting to the agenda of the Massachusetts SPCA. Rabbi Guinzburg saw no ambiguity in the position asserted by Angell. The idea voiced in the response to his first intervention in the issue of *Our Dumb Animals,* asserting that Jewish teaching commanding kindness and the absence of cruelty in the treatment of animals was not sufficient, was unequivocally nefarious. "Now this brands us altogether as a cruel people, which is indeed a grave charge; but the accusation does not rest here, for it implies a still graver

7 Rev. M.R. Miller, *The Luminous Unity* (Philadelphia, 1874), 260–263; "Aaron Guinzburg," in Miloslav Rechcigl, Jr., *Encyclopedia of Bohemian and Czech-American Biography,* Vol. 1 (Bloomington, IN: AuthorHouse, 2016).

8 "The Jewish Method of Slaughtering: The Rabbi's Answer," *Our Dumb Animals,* December 1870, 53.

9 Ibid., 52.

stigma on our religion, which, it is said, prescribes to us the practice of cruelty against defenseless animals."[10]

The rabbi returned to a spirited Biblical defense of Judaism and its treatment of animals as commanded by the "divine law-giver" and promised to reply specifically on the Jewish method of slaughter in the next issue. Rabbi Guinzburg finally addressed the practice of shechita for the readers of *Our Dumb Animals*. After a careful review of the practice and the various injunctions imposed on shochetim during slaughter, he concluded that the Jewish method was inherently and necessarily humane. He left critics of shechita in no doubt that he thought of them as people who "know very little of Judaism, theoretically or practically."[11]

The Massachusetts SPCA was not convinced by the rabbi's careful examination of the theology or practice of shechita. The front page of the same issue of *Our Dumb Animals* was filled with a complete description of discussions of the Jewish method of slaughter at the International Congress. The interventions at that meeting, as reported to readers in the United States, accepted that scientific opinion was divided on the cruelty of shechita in the actual process of slitting the animal's throat, with a single cut, although those who believed it to be cruel rejected the contrary medical opinion. The Congress condemned, without hesitation, the process of casting before killing as inherently cruel.[12] The editor of the magazine reasserted the SPCA's current state of neutrality on the question of the Jewish method of slaughter, an issue considered to be of broad public interest. "We devote so much space to it because it involves the whole question of slaughtering with the least pain. The public need enlightenment upon this subject, and the butchers can bear instruction.... We are not yet prepared to decide between them, but have deemed it our duty to let all sides be heard."[13]

These claims of neutrality were belied by the publication in the next issue of *Our Dumb Animals* of yet another full-page story covering the International Congress and giving voice to the majority of participants who opposed the Jewish method of slaughter. Over the objections of a minority of delegates who believed that interference with religious practice was outside the proper function of the Congress, John Colam, of the RSPCA and a close personal friend of George Angell, was reported to have argued that to any observer the

10 "A Defence of the Jews," January 1871, 57.

11 "The Jewish Manner of Slaughtering Animals," February 1871, 67.

12 "Slaughtering with the Least Pain to Animals," ibid., 65 and 72.

13 "The Jewish and Other Methods of Slaughtering," ibid., 68.

could perhaps be attributed to Jewish insistence on the Pentateuch and their refusal to recognize the universal humanity of the God of the New Testament. This idea and ideal of a humanity that rejected slaughter without stunning had been, would be, and continues to be, the operating rhetorical apparatus for the legal and moral attacks on the Jewish method of slaughter. A universal and Christianized humanity constantly and consistently does battle with Jewish religious practice. Animal welfare attitudes on the question of shechita by definition excluded observant Jews from the category of humanity itself.

After this flurry of activity around the shechita question, Angell focused his attention elsewhere. While the SPCA continued to monitor the transportation of animals for slaughter and the slaughtering process itself, and maintained its general humane education, Angell became an advocate for peace, for the end of vivisection, as well as for public health and hygiene. He specifically focused on issues of a safe food supply, free of diseased and dangerously adulterated products. In 1887, sixteen years after the debates over the Jewish method of slaughtering animals for food, Angell returned to the laws of kashrut. Instead of raising questions about the peculiar custom of observant Jews, one that gave rise to presumptions about its cruel nature, Angell turned his attention to the post-slaughter practice of carefully inspecting the meat of the animal to ensure that it was in no way diseased or otherwise unfit under Jewish law. While the theological underpinnings of his commentary might raise concerns, Angell offered a surprisingly positive assessment.

> The Jewish Rabbi goes to our markets and selects what seems a healthy animal. He stands at the slaughter-house while it is slaughtered and dressed. During the process, he carefully examines the internal organs, and if he finds the slightest disease, passes it over to the Christian.
>
> When public opinion shall demand the same inspection of animals, both before and after they are killed, now practiced in Continental European cities, and by the Jews, so far as I am informed, everywhere, and the Christian inspector shall stand at out cattle markets, side by side with the Jewish Rabbi, to condemn and cause to be destroyed the meat of every diseased animal, then animals will be brought to our markets without cruelty, and the Christian will eat as good meat as the Jew.[15]

15 George T. Angell, *Protection of Animals* (Boston, 1887), 12; "Jewish Slaughtering– Public Health," *Our Dumb Animals*, November 1894, 66.

Angell's analysis is grounded in an understanding of Jewish exceptionalism and an idea that public and Christian are largely, it not totally, synonymous. He does accept the validity of a key assertion made by supporters and defenders of shechita throughout attempts to criminalize or outlaw completely the Jewish method of slaughter on humanitarian grounds. He concludes that the application of the laws of kashrut with demands for healthy meat for human consumption was a practice that led to greater protection of public health. While at this point Angell remained silent on the question of shechita and cruelty, he made it clear to his audience that there was much to be admired and replicated in Jewish religious practice and its concern for the provision of healthy meat to consumers. For the remainder of his time as the head of the Massachusetts Society for the Prevention of Cruelty to Animals, there is no real indication that Angell or his organization focused on shechita. However, when he died in March 1909, a different and more perniciously antisemitic agenda quickly emerged in the commonwealth.

Led by Francis H. Rowley, a well-known Baptist minister, who served as its president from 1910–1945, the Massachusetts organization campaigned to outlaw shechita in that state.[16] Rowley had been Secretary for the SPCA under Angell and succeeded the attorney as President. Under his leadership, the organization fought more vigorously and publicly for humane slaughter. "Probably Dr. Rowley has done more than any other individual in this country towards creating a public opinion in favor of humane methods of slaughtering cattle, swine, and sheep."[17]

As in many other instances of SPCAs and their campaigns against shechita, arguments were placed into this category of humane slaughter. More often than not, they were accompanied by declarations of an absence of any antisemitic motivation. Rowley and the Massachusetts SPCA followed this general pattern, but there is no real difficulty in uncovering an unambiguous targeting of the Jewish method of slaughter. That targeting, within a narrative that juxtaposed humanity and Jews, and that consistently invoked tropes of blood to condemn shechita, evidences antisemitic effect and what many would see as deep and persistent evidence of antisemitic intent. Despite his protestations to the contrary, Rowley himself did not hesitate to invoke clear metaphors and theological critique that could leave little doubt about

16 "Rev. F.H. Rowley of S.P.C.A. Unit Dies," *New York Times*, February 15, 1952.

17 Coleman, *Humane Society Leaders in America*, 111–112; Francis H. Rowley, *The Humane Idea* (Boston: American Humane Education Society, 1912).

his understandings of Jewish religious practice and his attitude about the defense of shechita by leading rabbinical authorities:

> It may be said, however, that many hold—they believe with entire reason—that for that appalling institution of bloodshed which was a part of Israelitish worship there was no actual divine sanction; that we must see here the persistence of traditional customs and the influence of a priestly order with ends to subserve that were widely different from those great moral goals which rose before the inspired visions of the prophets.[18]

The Baptist minister presumes to attack shechita not only as an "appalling institution," but as one that is not the result of divine command. Some sort of rabbinical conspiracy, pursuing unarticulated, but pernicious, and other than theologically sound ends, is responsible for preventing a more authentic Judaic practice that would, by happy coincidence, directly mirror New Testament humanity. Given the cruel nature of the Jewish method of slaughter, and its inauthentic nature, the campaign against inhumane slaughter could and must, as a moral imperative, seek to ban shechita. "Here is the supreme task and duty of the American humane societies of our time. We are as far behind such a country as Germany in this respect as barbarism is behind civilization."[19]

Rowley voiced views that had been informing the work of the SPCA in many jurisdictions for several years.[20] As in New York under Bergh, the Massachusetts Society had pursued perpetrators of acts of animal cruelty through its police powers of investigation, inspection, and prosecution. The archival records are sparse, and there are no official minutes or reports detailing the particular instances.[21] There is some evidence that prosecutions involving aspects of shechita had been brought, however sporadically, by SPCA inspectors. *Our Dumb Animals*, for example, reports that in 1895, "for barbarously torturing a steer (by hanging him up etc.,) a Jewish butcher was fined $10."[22]

18 *The Humane Idea*, 5.

19 Ibid., 56.

20 "Is the Jewish Mode of Killing Oxen Enforced by Scripture?" *Animal World*, January 1882, 26.

21 Jan Holmquist, Assistant to the President, Massachusetts SPCA, Personal Correspondence, November 20–21, 2016.

22 "From Report of Charles A. Currier, Chief Prosecuting Agent, of Complaints Investigated from March 1, 1894, to March 1, 1895," May, 1895, 133.

It is impossible to glean from this account enough detail to allow us to assert with certainty that this is a shechita case. The convicted person was a "Jewish butcher," who may or may not have been qualified as a shochet. Beyond a perhaps pervasive antisemitic atmosphere, the specific characterization of the individual as a Jewish butcher might serve to indicate that there was something Jewish about the slaughtering process. The description of the case as one involving "hanging him up" might indicate that the animal was cast as a part of shechita. Finally, there is no way of knowing what the "etc." embodies. Is this meant to include the rest of the casting, if it is casting, or does "etc." include the slitting of the animal's throat without stunning, thereby turning this into an early example of an American shechita prosecution?

The following year, a similarly incomplete summary of a prosecution in Massachusetts raises more intriguing, vexing, and currently unanswerable questions. "A Jew butcher, who thrust his fingers into the eye of a steer he was slaughtering, was fined $12."[23] The choice of the descriptor "Jew butcher" can be seen as a manifestation of an unabashed antisemitism or perhaps as signifying a shochet engaged in the Jewish method of slaughtering. The allegation of those involved in shechita thrusting into the eye of the animal about to be slaughtered is a commonplace in cruelty allegations and prosecutions. The state of the archival record leaves this instance as another isolated case in the anti-cruelty campaign of the Massachusetts SPCA. Neither of these two instances appears to have entered into the public discussion of slaughterhouse cruelty or into the SPCA narrative of specific attacks on the Jewish method of slaughtering. In each case, the Jew was convicted of cruelty, and one might have expected the Society to make much of concrete examples of cruelty in its campaign. Perhaps it is possible to suggest that they were not real shechita prosecutions, since anti-cruelty authorities would otherwise have invoked them as concrete examples of the inhumane nature of the Jewish method being carried out in the abattoirs of the state. The absence of any evidence that the cases became "major attacks on Jewish freedoms" might indicate that they were simply instances of real cruelty to animals during slaughtering, and had nothing to do with shechita. Given the state of the available records, we probably will never know the answer to these questions.

---

23 "From Report of Charles A. Currier, Chief Prosecuting Agent, of Complaints Investigated from March 1, 1895, to March 1, 1896," May, 1896, 151.

Some years passed, during which the Massachusetts SPCA continued to agitate for slaughterhouse reform. Two years after the New York SPCA mounted another campaign for humane slaughter, raising fears in the New York Jewish communities that they were about to live through another sustained attack on shechita, animal advocates in Massachusetts began an assault on the Jewish mode of slaughter.[24] In December 1910, Rowley had called on readers of *Our Dumb Animals* to ensure that legislatures quickly adopted new regulations on public abattoirs throughout the country. He urged his readers to support calls for banning the Jewish method. "Give us the public sentiment and we can obtain the law and enforce it. Even the Jewish method of slaughter must soon yield to a more enlightened conception of what is humane. Some one has said that no religion can be acceptable to God which compels its devotees to be cruel to men. He might justly have added also, 'cruel to any living creature.'"[25]

The opposition between the humane and the Jewish is buttressed by a directly theological invocation of the one true faith and its precepts, as Rowley's underpinning Protestant antisemitism became unrelenting. But as 1910 came to an end, the Massachusetts SPCA would turn to a more prosaic mechanism in its campaign to end shechita, or so it would appear. Acting on orders from Rowley, inspectors over a seven-week period investigated hygienic and slaughtering concerns in a number of facilities throughout the commonwealth, particularly "in many suburban places." In late December,

> ... warrants were issued for the men in a slaughter house in Attleboro, conducted by Samuel Alpert. These were Rabbi Katz Nelson, G.E. Dawson and Samuel Adelson, and all gave bail. The agents allege that in this case cattle were being killed without an inspector being present, and that the animals were being treated cruelly. It is alleged that calves less than nine days old were killed and that some of the meat from the slaughter house is sent to Rhode Island.[26]

24 "Jewish Slaughtering and Humanity," *American Hebrew and Jewish Messenger*, March 6, 1908; "A New Assault on Shechitah," and "Reported Attack on New York Shechitah," ibid., June 19, 1908.

25 December 1910, 104.

26 "As to Diseased Cattle: Agents of S.P.C.A. Are Instructed to Proceed Against All Places Where Laws Are Violated," *Boston Daily Globe*, December 22, 1910.

The Attleboro Shechita Case: Small Town Jewry and the SPCA

Much of the activity of the Massachusetts SPCA surrounding the shechita question was centered in Boston, a city with a long history of Jewish presence, dating from the earliest colonial period.[27] Consistent with dominant trends in Jewish population dynamics, the periods of increased immigration in the mid- to late nineteenth and early twentieth centuries, the largest number of newly arrived Jews settled in metropolitan centers. By the second decade of the twentieth century, Boston's Jewish population numbered 77,500, almost ten percent of its overall 775,000 inhabitants.[28]

The court case in which it appears that the Jewish method of slaughter was condemned as cruel took place not in Boston, but in the town of Attleboro. Attleboro is situated in southeastern Massachusetts, some forty miles from Boston. Attleboro is much closer to the state of Rhode Island, and is a little over twenty miles from Providence. It is unsurprising that the "Attleboro shechita case" should involve the supply of meat to Jews in Rhode Island. Recent immigration from "Russia" had raised the Jewish population in Providence to 15,000.[29] Attleboro at the time just after the shechita case had a Jewish population of only 530.[30] It was only at the time the case was proceeding through the courts that the first Jewish congregation in Attleboro, Agudas Achim, had been formed when seven local men purchased the former premises of the Swedish Congregational Church to serve as the synagogue.[31]

The Attleboro shechita case occurred away from major metropolitan centers, in small town America, a context that Lee Shai Weissbach has argued is missing from many historical accounts of American Jewry.[32] So-called small town life has come to play a central role in the historical and mythological construction of American existence, and placing local Jewries within this context allows us to tell more complete stories about United States Jewish history and the ways in which the multi-layered aspects of life in small-town America

---

27 Jonathan Sarna, Ellen Smith, and Scott-Martin Kosofsky, *The Jews of Boston* (New Haven, CT: Yale University Press, 2005); William Braverman, "The ascent of Boston's Jews, 1630–1918," PhD dissertation, Harvard University, 1990.

28 Samson D. Oppenheim, "The Jewish Population of the United States," *American Jewish Yearbook*, 20 (1918–1919): Table A, 49.

29 Ibid., Table B, 50.

30 Ibid., Table F, 55; Table H, 60.

31 Laura Calverley, "Religion Played Major Role in Attleboro's Development," *Sun Chronicle* (Attleboro), July 6, 2014; "Jottings," *American Israelite*, August 17, 1911.

32 *Jewish Life in Small-Town America* (New Haven, CT: Yale University Press, 2005).

interacted. Another part of American Jewish existence in the late nineteenth and early twentieth centuries sits at the periphery of the Attleboro case, since the area around the city and towards the Rhode Island border was home to many Jewish farmers, who would, in all likelihood, have provided some of the animals to be slaughtered at the Alpert abattoir.

There is a long history of Jewish agricultural communities in the United States.[33] A number of Jewish farmers were established in the area between Attleboro and Rhode Island in 1910 and 1911. Jewish farmers in the Attleboro area were active and avid participants in the short-lived Federation of Jewish Farmers of America.[34] Many of these Jewish farmers around Attleboro pursued more than agricultural life. They organized, together with their urban counterparts in the town, in order to carry out, as far as possible, authentically Jewish lives. One farmer, Charles Fine, a former yeshiva student in his native Russia, left New York, his first home in the United States when he emigrated in 1900, and settled on a farm outside Attleboro in 1904. Soon after his arrival, he organized a *minyan*, and acted as an unofficial rabbi for the small community, a not unusual occurrence in less populous areas. Like the circuit riding rabbis in small town Nova Scotia, he officiated at weddings and funerals, and acted as shochet, as well as leading Sabbath services in premises rented in Attleboro, all this before the formal establishment of Agudas Achim.[35]

Like Jews in colonial America, and in other smaller communities in the Mid-West and the western United States, Jews in New England often struggled to follow the rules of kashrut. As Weissbach establishes, this must itself be placed in different regional and historical contexts. The Spanish and Portuguese populations of New York, for example, employed shochetim from their earliest days in the city. Likewise, the history of the Sephardic community in pre-Revolutionary Newport, Rhode Island, establishes the presence of a shochet.[36] Later immigration, to different parts of the United States, by so-called German Jews, many of whom followed Reform Judaism, meant that

33 Irwin Weintraub, *Jews in American Agriculture: An Annotated Bibliography* (Jefferson NC: McFarland, 2007).

34 "Attleboro," *Jewish Advocate,* March 16, 1916; "Jottings," *American Israelite,* July 10, 1913.

35 Eleanor F. Horvitz and Geraldine S. Foster, "Jewish Farmers in Rhode Island and Nearby Massachusetts," *Rhode Island Jewish Historical Notes* 10, no. 4 (1990): 445–446, 470.

36 Holly Snyder, "Reconstructing the Lives of Newport's Hidden Jews, 1740–1790," in *The Jews of Rhode Island,* eds. George M. Goodwin and Ellen Smith (Hanover, NH: University Press of New England, 2004), 32–33.

> ...in their desire to conform to the culture of the majority, even those who maintained their identity as part of a religious community quickly gave up the observance of the Jewish dietary laws, for example. While the abandonment of *kashrut* was a general trend among Reform Jews, it may have proceeded even more rapidly in smaller Jewish settlements, given the greater difficulty of obtaining kosher foods in those places.[37]

In a larger center such as Providence, Jewish communities tended to form synagogues not along the lines of Orthodox versus Reform, but according to local identities from the old country. "By 1914, immigrants had organized twenty-four different religious congregations, among them several different Austrian and Roumanian groups."[38]

Many of the eastern and central European Jews who arrived in the United States at the end of the nineteenth and beginning of the twentieth centuries, who constituted the populations of Attleboro and Providence, were Orthodox in their religious practices, and in particular in their attempts to comply with the laws of kashrut, a key part of which would have been the consumption of meat from animals killed according to the practice of shechita.[39] The earliest Jewish immigrants in Providence, regardless of their country or region of origin, attempted to comply with the laws of kashrut. Berman records the presence of a shochet in the city in the late 1850s.[40] Throughout rural New England, in places like Attleboro, Jews would have sought to engage the services of a shochet, either as a permanent fixture, or to use travelling rabbis.[41] The Jews of Attleboro were recent immigrants to the United States at the time of the SPCA prosecution. They had only begun the process of organizing into a formal religious community with a synagogue. They were small businessmen or farmers, and they were religiously observant. The practice of shechita would have been an essential part of adhering to the laws of kashrut and living their lives as good Jews, while at the same time being good Americans.

Whatever differences existed between and among the different synagogues and their members in Providence, they were religiously observant

---

37 Weissbach, *Jewish Life in Small-Town America*, 227.

38 Judith E. Smith, *Family Connections: A History of Italian and Jewish Immigrant Lives in Providence, Rhode Island, 1900–1940* (Albany, NY: SUNY Press, 1985), 136.

39 Ibid., 249–252.

40 *Shehitah*, 357.

41 Michael Hoberman, *How Strange It Seems: The Cultural Life of Jews in Small-Town New England* (Amherst: University of Massachusetts Press, 2008), 220–233.

Jews, for whom a supply of kosher meat would have been of great significance. The Jews of Attleboro and Providence ensured that they had access to kosher foods and to the services of shochetim. With the many farms on the Massachusetts side of the border, and the growing population of observant Jews in Providence, it should not be surprising that "rabbi Katz Nelson" would have come to the Alpert slaughter house in Attleboro in search of meat prepared according to the laws of kashrut. But in 1910 and 1911, Samuel Alpert and "Katz Nelson" stood accused of offenses relating to animal cruelty for acts carried out at the Massachusetts slaughterhouse.

The Attleboro Shechita Case and More Uncertainties in Massachusetts Jewish Legal History

The prosecution brought by the Massachusetts SPCA against the four men in the Attleboro slaughterhouse at the end of 1910 is characterized by two fundamental uncertainties, each of which is the result of the paucity and lacunae of the primary archival records and of the inaccuracies of contemporary newspaper coverage. The identities of the accused are error prone and the fundamental question of whether the cruelty charges related to the Jewish method of slaughter remains obscured in some accounts. The Jewish newspaper coverage of the case was dependent on reports in the non-Jewish press. Errors, particularly in relation to the identity of the parties and the lack of detail about whether the case was one involving the Jewish method of slaughter abound in that coverage. With the caveats informed by the poor state of the historical record, the prosecution of the four men in the Attleboro abattoir fits into the broader context of the SPCA's struggle against inhumane and unsanitary conditions in slaughterhouses in the Commonwealth of Massachusetts generally, and more specifically into its battle to end the Jewish method of slaughter.[42]

The full judicial files relating to the case can no longer be found, and all that remains are a few three- by five-inch index cards recording the names of the accused and the final disposition in relation to each.[43] Even in these records, one obvious error as to the name of the accused is present.

Temporally, it fits with Rowley's instructions to his inspectors to increase their vigilance and to bring prosecutions against acts of cruelty, particularly in

42 "Public Abattoirs," *Our Dumb Animals*, October 1910, 72.

43 James P. Perkoski, First Assistant Clerk-Magistrate, Attleboro District Court, personal communication, March 10, 2017; Debra A. Gardella, Fall River Superior Court, personal communication, February 1, 2017.

some suburban slaughter houses. From the historical records of Massachusetts and Rhode Island Jewries three of the accused, "Katz Nelson," Samuel Alpert, and Samuel Adelson were Jewish. When three Jews are arrested, and charged with animal cruelty by an inspector of the Massachusetts SPCA in an abattoir in suburban Massachusetts, there is, presumptively at least, a suspicion that this is a prosecution related to shechita.

Evidence of the identities of some of the accused lends support to this view. There is little, if anything in the records in either Massachusetts or Rhode Island that permits a convincing identification of George Dawson. Likewise, Samuel Adelson does not figure in census, residential, or business records in Attleboro or surrounds around 1910. Samuel Alpert and "Katz Nelson" are, to different degrees, more easily identifiable and their Jewish identities more readily confirmed.

The 1910 Census records that Samuel Alpert, a Yiddish speaker, resided in Attleboro, and was a butcher in a slaughter-house.[44] The 1909 Attleboro business Directory identifies him as a cattle dealer.[45] The first newspaper account of the prosecution identifies Alpert's slaughterhouse as the place where the alleged offenses occurred.[46] The 1910 prosecution of the four men in Attleboro on charges of animal cruelty clearly took place at the abattoir run by a member of the town's Jewish community.

The identity of the rabbi charged in the case is more problematic. None of the Massachusetts or the Rhode Island census, business, and other directories from either state, identifies a "Katz Nelson" or a "Nelson Katz" residing there at the time. Lists of clergy do not include a rabbi under either name. These records do identify a Wolf Katznelson, sometimes described as a rabbi and who is, given the circumstantial officially recorded evidence, almost certainly the "Rabbi Katz Nelson" described as one of the accused.

Wolf Katznelson, a naturalized citizen, had lived in Providence since 1895. Census records and city directories from 1900 to 1915 consistently identify him as a minister, a reverend, or a rabbi.[47] The 1900 *City Directory* enters his name under Grocers—Beef Dealers.[48] In 1915, he served on the Finance and

---

44 Thirteenth Census of the United States, 1910, "Attleborough Town," June 14, 1910.

45 *Attleboro Directory*, 1909, 35.

46 "As to Diseased Cattle," *Boston Daily Globe*.

47 Twelfth Census of the United States, 1900, "Providence City," June 15, 1900; *Providence Directory*, 1908; 1910, 372; 1911, 390; 1914, 778; Rhode Island State Census, 1915, "Providence City, Ward 3."

48 242.

Jubilee Committee of the Providence synagogue Sons of Zion (Bnai Zion).[49] The *Providence Directory* of 1918 shows an entry for Wolf Katznelson as a butcher living at 82 Orms Street, in a Jewish section of the city.[50] By 1922, Katznelson, living at the same address, is described as an inspector.[51]

The documentary trail leaves little doubt that "Katz Nelson," the rabbi charged with animal cruelty offenses at Alpert's Attleboro abattoir in December 1910 was Wolf Katznelson, rabbi, beef dealer, butcher, and inspector of Providence. It would not be unreasonable to speculate that a "rabbi, beef dealer, butcher, inspector" was a shochet. As Segal recounts, one of the earliest leaders of the congregation, Reb Eliezer Lipshitz, who served as *hazan* and shochet, "kept a meat market at 35 Mill Street" (listed in the directory as dealing in "provisions").[52]

The prosecution was brought not by a local Attleboro representative of the SPCA, but by Henry Perry, an agent of the Massachusetts SPCA, based in Mansfield, some seven miles distant.[53] Perry charged the accused with slaughtering outside the presence of an authorized inspector and attempted to prosecute local inspector Caleb E. Parmenter for failing to be present at the killing of the cow by Katznelson. The SPCA agent argued that because Parmenter left the abattoir before the cow was killed, he was guilty of failing to perform his legal duty, and was as liable as the butchers themselves. The court found on the facts that Parmenter had been present up to a few minutes before the beast was killed and that he had given permission for the slaughtering to take place, thus fulfilling his legal obligations.[54] Perry's zealous enforcement of the law and of Rowley's slaughterhouse campaign extended beyond the Jewish method, but it did focus on the killing by Katznelson.

The Boston-based *Jewish Advocate* offered a reasonably full retelling of the conviction of the four men at the District Court in Attleboro in December 1910. "Rabbi Nelson" was fined one hundred dollars; Adelson, the same amount; Alpert, seventy-five dollars, and Dawson, fifty dollars. The description of the case presented by Agent Perry and accepted by the Court leads not

49 Beryl Segal, "Congregation Sons of Zion Providence Rhode Island: The Orms Street Synagogue," *Rhode Island Jewish Historical Notes* 4, no. 3 (1965): 276.

50 433.

51 *Providence Directory*, 1922, 479.

52 Segal, "Congregation Sons of Zion," 241.

53 *Mansfield Directory*, 1911, 88, 122.

54 "Cattle Inspector Not Guilty of Failing to Oversee Slaughtering," *Attleboro Sun*, December 23, 1910.

illogically to the idea that this case did involve the Jewish method of slaughtering animals for food. Perry

> ... alleged that Rabbi Nelson and two assistants, Dawson and Adelson, were killing a cow at Alpert's place. The animal was hauled up by a rope attached to one leg and its weight was thrown on to its forward legs. Adelson struck the animal in the legs with a shovel, inflicting several wounds. When the agent protested, the Rabbi is said to have directed Adelson to go ahead with the work.[55]

This appears to be a description of the process of casting the animal in preparation for the shochet's blade. Katznelson, the rabbi, was in charge of the preparation of the cow for killing. He directed Adelson's work. The beating and wounding of the animal in the process of casting violates general Jewish precepts that kindness to animals is an overarching ethical obligation incumbent at all times. This may be a case in which there was real cruelty to an animal during preparation for slaughter contrary both to Jewish law and ethics and to the legal regime of the Commonwealth of Massachusetts. This makes the proper characterization of the case not just speculative, but increasingly problematic. It may not be a shechita case because the ethics of human relations with animals in Jewish law prohibit acts of wanton cruelty.

The four convicted men mounted an appeal and the case was reheard before the Massachusetts Superior Court in New Bedford in June of 1911. The *Providence Journal* offered a brief summary of the decision. The men were again convicted and each fined fifty dollars "for cruelty in the method of slaughtering cattle at Alpert's slaughter house on Willmarth Street, Attleboro." The Rhode Island paper's account continued that "this case is considered important, the dispatch continues, by the Society for the Prevention of Cruelty to Animals as it prevents officials of the Hebrew church from using undue cruelty in the killing of animals for Kosher meat."[56]

The ascription of the importance of the case to the SPCA offers support for understanding this case as an anti-shechita prosecution. It fits with the broader SPCA campaign against local slaughterhouses and against the Jewish method. It is consistent with Rowley's long-standing opposition to shechita and his ongoing attempts to end the practice in Massachusetts. The *Providence Journal*

55 "Rabbi Fined for Cruelty: Three Other Men Accused in Court at Attleboro," December 30, 1910.

56 "Providence Rabbi Fined," *Providence Journal*, June 24, 1911.

reported that the animal welfare organization viewed the case as one that confirmed its opposition to the use of "undue cruelty in the killing of animals for Kosher meat." As the *American Israelite* underlined, "the paper does not state whether the men were unnecessarily cruel or whether it was the Jewish method of slaughtering for which they were fined."[57]

Some of the ambiguity surrounding the case is clarified in the reporting in the local newspaper, the *Attleboro Sun*. In its account of the original trial of the four men before the District Court, the paper repeated the allegations of SPCA Agent Perry that the animal, at the insistence of Rabbi "Nelson," was struck on its legs with a shovel in order to make the animal fall on to its side. Once the animal was in this position, "the rabbi cut its throat with a long knife," after which "the animal was then allowed to bleed to death, this requirement being necessary in the Jewish religion.[58]

The account, like most others in relation to the case, is not without its inaccuracies, either stemming from the reporter's own errors, or from the incorrect assertions of the SPCA representative. The error on the shochet's name is consistent throughout the retelling of the Attleboro anti-shechita case, Katz Nelson for Katznelson. Agent Perry alleged that the animal was allowed to bleed to death, implying that the cow was still alive after Katznelson had delivered the cut with his "long knife." If the SPCA agent's account is accurate, then at least two possible instances of unnecessary cruelty have been catalogued. There is the blow to the animal's legs in a crude casting process, with resultant injuries. The second instance would be in allowing the animal to experience pain as its was bled to death in fulfilling a "requirement ... necessary to the Jewish religion."

Consistent with other instances throughout the SPCA's attempts to criminalize shechita, it is possible that Agent Perry misunderstood not just Jewish religious law, but the science of the process of slaughter. Once a successful cut has been made by the shochet, blood flow to the brain has stopped, and the animal's central nervous system has been severed. The animal is insensate and under many definitions, dead. Blood is removed from the animal according to the laws of kashrut. It is inaccurate both physiologically and theologically to insist that the animal is bled to death in the way this phrase seems to have been understood by the SPCA agent.

---

57 "Jottings," July 13, 1911.

58 "$325 Fines on Rabbi and Three Others; One is Again Arrested," *Attleboro Sun*, December 22, 1910.

The description offered by Perry and reported in the *Attleboro Sun* is that "Nelson" was in charge of preparing the animal for slaughter, and that he killed the cow with his long knife. What the SPCA agent recounts is the Jewish method of killing an animal for human consumption, completed by a shochet employing the special blade or *hallaf*. The local newspaper account transforms the case from an instance with vague assertions of unnecessary cruelty, into one in which the Jewish method is an essential element. The *Attleboro Sun* story adds that a leading, but unidentified, rabbi from Providence gave evidence as to the practice of shechita, following some controversy as to whether he would be permitted to swear the oath and to give his testimony with his head covered, "in conformity with his religion."

Perry denied the anti-Jewish element, claiming that he did not object to the method of killing, but did not accept the cruelty that accompanied the process. Photographic evidence of a dead cow hanging by its hindquarters was admitted to confirm the SPCA's case about the generally cruel conditions obtaining in the slaughterhouse. After hearing the SPCA agent, the four accused, and the religious expert, the judge convicted the men.

The religious element of the prosecution is inescapable. This is a case where a central focus was on the role played by shochet Katznelson in commanding the process, and in slitting the animal's throat. The idea that the animal was bled to death because of the requirements of Jewish religious law was clearly in the mind of the SPCA agent. Combined with the presence of a religious expert witness, these elements tend to establish that this was an anti-shechita prosecution, despite the claims of Agent Perry that he had no objections to the manner of killing.

The potential underlying ambiguities of the case were in evidence at the hearing of the appeal. The District Attorney, acting as the prosecutor before the Superior Court, opened his address to the jury by stating that this was not a case about religious belief and practice. The Jewish method of killing animals for food could be, and was, carried out in a manner that was not unnecessarily cruel. What was before the jury was the question of how "Rabbi Nelson" performed his tasks on a particular day in Attleboro, not shechita itself.[59]

But the case, despite Agent Perry's and the District Attorney's claims to the contrary, remained inextricably a Jewish case, an anti-shechita prosecution. Much of the testimony Perry had given in the Attleboro District Court trial was repeated. Evidence from local law enforcement officials who had attended

59 "Rabbi's Method of Killing Beef Tested in Appealed Case," *Attleboro Sun*, June 21, 1911.

the slaughterhouse after the cow was killed was offered to demonstrate that the animal had been lifted by its hind legs tied with rope and bled. A statement proffered by George Dawson, described as a local butcher, and one of the co-accused, to Agent Perry confirms the characterization. Dawson was alleged to have told the SPCA agent at the time that "This isn't my way of killing, but I'm hired by these people and I have to do it their way."

"These people," Adelson, Alpert, and "Nelson", are the three Jews in the case, and "their way" can only really be the Jewish way. For Dawson, the non-Jewish butcher, and for the SPCA agent Perry who offered his words as part of the evidence for the prosecution, the case that was not one involving any attack on religious belief, was precisely one that targeted the cruelty of a method of killing animals for food pursuant to the religious belief and practice of "these people." Other witnesses contrasted the Jewish method of casting with the common practice of stunning. During the testimony of "Katz Nelson," who claimed that he simply fulfilled his duty under Jewish law by killing the animal with a single, continuous cut, the judge intervened to ask if the method described by the rabbi could be found "in the Christian Bible."[60] What the *Attleboro Sun* called the "kosher meat" cases, cases that for the SPCA and the District Attorney, did not concern religious belief, but in which the "rabbi's method of killing beef" was tested in law, and contrasted with killing with stunning, resulted in the conviction of "Katz Nelson" on charges of animal cruelty.[61]

Even after the legal process of an appeal had been completed, some ambiguity of the Attleboro shechita case remains. From the description of Perry's evidence at trial we might conclude that the men were unnecessarily cruel in casting the animal and that was the basis on which they were convicted. If that were the limit of the case, both Jews and the SPCA would be in agreement that the convictions were justified because they both abjure cruelty to animals. But if this was an attack on shechita itself, the familiar debates, disagreements, and pernicious motivations surrounding attempts to criminalize the Jewish method of slaughter would rise to the surface. In their summing up, the attorneys for the defense and prosecution put the question of the SPCA's and Perry's motivation at the forefront of their arguments. The defense lawyer claimed that Perry had a particular agenda and that his testimony had been tailored and exaggerated in order to obtain a conviction of the Jewish defendants. The District Attorney

60 "Arguments Heard in Kosher Meat Cases," *Attleboro Sun*, June 22, 1911.

61 "Fines in Kosher Meat Cases $225 But Less than in Lower Court," *Attleboro Sun*, June 23, 1911.

likewise told the jury that the motives of the SPCA in bringing the case were clearly at the heart of their deliberations.

The few remaining judicial files, in the form of the index cards recording the facts of the outcome in New Bedford, do not really shed light on the key taxonomical questions, but they do give rise to some further questions. George E. Dawson, the non-Jewish defendant, was fined twenty-five dollars for "Cruelty to Animals."[62] Samuel Adelson, the Jew, was fined seventy-five dollars for the same offense.[63] Samuel Alpert, the owner of the abattoir, was fined fifty dollars for the separate offense of "killing calf under 4 weeks old."[64] On the animal cruelty charge, however, Alpert was found not guilty.[65] Reports that the four were convicted of cruelty in the process of killing animals in the slaughterhouse in Attleboro are not, strictly speaking, accurate. The owner, Alpert, was acquitted on cruelty charges and only convicted on the separate offense relating to killing young calves. This crime was a per se offense, meaning that it was a crime to kill a calf younger than four weeks in any circumstances. The offense did not carry with it any idea or requirement of cruel treatment. Witness accounts contain no mention of Alpert's participation in casting or in the killing of the animal in which the cruelty is alleged to have occurred and continued under the rabbi's direction. He appears to have been charged as the owner of the slaughterhouse, or as Adelson's employer. It is not surprising that if that were the case, he would be acquitted on charges relating to events at which he was not present. As far as "Nelson Katz" was concerned, the court documents demonstrate that the state did not proceed against him on the charge of killing a young calf and a notice of nollei prosequi was recorded.[66] He was however convicted of "Cruelty to Animal" and fined fifty dollars.[67] Each accused paid his fine to the sheriff on the same day.

We can never know if the jury returned the verdicts in this case because they accepted the testimony of the SPCA agent and confirmed his good faith in attempting to prosecute an individualized instance of animal cruelty in the Attleboro abattoir, or if they rendered their verdict because they shared in a religious prejudice against "these people." The lowest fine was imposed on

62 New Bedford Superior Court, Docket No. 3096, June 22, 1911.
63 Docket No. 3099, June 22, 1911.
64 Docket No. 3100, June 22, 1911.
65 Docket No. 3098, June 22, 1911.
66 Docket No. 3101, June 22, 1911.
67 Docket No. 3097, June 22, 1911.

the non-Jewish butcher, and the heaviest on the Jewish Adelson. According to Perry's evidence, it was Adelson who carried out the casting and inflicted the injuries on the animal. Katznelson, who directed the operation according to Perry, received a fine half-way between the two butchers.

The preponderance of circumstance shows that this was a case about "kosher meat" and that it involved allegations of animal cruelty against a shochet, about the "Rabbi's method of killing beef." There are remaining uncertainties in relation to the facts of the case, issues of casting, as well as questions of the killing process itself. Was this a poorly executed killing of the cow, involving a violation of both religious and secular prohibitions on cruelty, or was this a normal act of shechita criminalized by the SPCA and the judicial system of Massachusetts? Definitive answers elude us, but in the context of the SPCA's attacks on shechita in Massachusetts at the time, and in light of the characterization of the trial in the press as the "kosher meat case," accompanied by a judicial intervention about the Jewish method of slaughter and the Christian Bible during Rabbi Katznelson's testimony, evidence by witnesses about the normal use of stunning before killing, and expert testimony from a Providence rabbi, understanding this case as an instance of anti-shechita prosecution and persecution is a logical conclusion.

While the account in the *Providence Journal*, replicated in the *American Israelite*, indicated that the case was of significant import to the SPCA in its campaign to end shechita, there does not seem to be any evidence that the conviction figured prominently in the SPCA's ongoing battles to impose humane slaughtering in Massachusetts. There appear to have been no further prosecutions in the state in relation to the Jewish method of slaughter. The convictions were not the object of comment, concern, or intervention at the executive level in Boston.[68] Details of the case, a successful prosecution of a "Jewish rabbi" involved in cruelty during the slaughter of a cow for food, were not passed on to Robert Murray in Halifax during the prosecution of Abraham Levitt a mere two years after the Attleboro and New Bedford judicial proceedings. The only indication that the Attleboro shechita case figured in any way for the Massachusetts SPCA is a brief mention in its official journal. "Three Jewish butchers and a rabbi of Attleborough were found guilty of cruelty in slaughtering and fined in the lower court, from which they appealed. The case

68 Jan Holmquist, Massachusetts SPCA, personal communication, March and May 2017.

came up in the superior court in New Bedford when the jury returned a verdict of guilty. Fines aggregating $225 were paid by the four offenders."[69]

Overlooking the continuing inaccuracies in press accounts of the case, e.g. Katznelson was from Providence not Attleboro; Dawson was not Jewish; and one of the Jewish butchers, Alpert, was not convicted of cruelty, it is still remarkable that the SPCA had clear knowledge of the case, had prosecuted and achieved the convictions, maintained on appeal, of two Jewish butchers and a rabbi/shochet, yet no more was made of these convictions. The Jewish communities of Massachusetts and Rhode Island, like the SPCA, appear to have dealt with this case by ignoring it. Unlike cases in the United Kingdom, Australia, and Canada, or in New York and Denver, where prosecutions or threats of prosecution of shochetim were considered to be "major attacks on Jewish freedoms," this case, which resulted in convictions and fines, seems to have passed under the radar and into oblivion.

Perhaps there was such Jewish silence because there was cruelty in this particular instance that violated both secular and Jewish law. But some intriguing facts do remain and justify further exploration. The case was obviously well-known to the Jewish communities, since "the chief rabbi" of Providence had testified at the Attleboro trial.[70] Each of the defendants paid the fine to the sheriff on the day the verdicts were entered in New Bedford. Two hundred twenty-five dollars in 1911 was the equivalent of over six thousand dollars in today's currency. One might inquire where these men, none of whom appears to have been particularly well-to-do, obtained the money to pay the fines. This case and these convictions occurred at a time when the Massachusetts SPCA under the active leadership of Francis Rowley, a vocal opponent of shechita, was in the middle of a public campaign that placed the Jewish method of slaughter at the forefront of political and legal debate. It pitted the SPCA against an increasingly well-organized and vocal opposition from the Jewish communities of Massachusetts. The disappearance of the Attleboro shechita case from legal history, Jewish and anti-cruelty memory, is puzzling to say the least.

The SPCA, Shechita, and Massachusetts Politics

In the months following the successful prosecution of the Attleboro four, and immediately preceding the conviction of the Jewish butchers before the

69 "Exceptional Cases," *Our Dumb Animals*, August 1911, 41.
70 "$325 Fines on Rabbi and Three Others," *Attleboro Sun*, December 22, 1910.

Superior Court in New Bedford, the SPCA had attempted to introduce a legislative ban on the Jewish method of slaughter. The SPCA sought to eliminate slaughter without stunning through an act of the state legislature. Elsewhere, the SPCAs had proceeded by way of an individual criminal prosecution, hoping that a conviction would serve as a springboard to broader prohibitions on slaughter without stunning. The Attleboro case, and the other scattered instances from the records of the Massachusetts SPCA, indicate that the organization did pursue criminal prosecutions of Jewish butchers for animal cruelty and that they had some success. But for Rev. Dr. Rowley, the aim was not just to prosecute individual acts of cruelty, but to put an end to the practice of killing without stunning, and as a consequence to place the anti-shechita campaign within a broader context of radical changes in the operations of abattoirs in the commonwealth.[71] The goal was to eliminate cruelty in animal slaughter in the name of humanity.

The broader campaign would have obvious antisemitic goals and effects, and perhaps motivations to which the label could be attached. If successful, the introduction of compulsory stunning would end the Jewish method. Given the underlying narratives contrasting humane slaughtering practices and animal cruelty, the introduction of the proposed SPCA ban would characterize Jewish religious practice as fundamentally and inescapably inhumane and inherently cruel.[72] The SPCA introduced general legislation prohibiting slaughter for human consumption without prior stunning.[73] In section 2 of the bill, it had sought to provide that

> ... no animal in the process of slaughtering shall be caught up or hung by its legs ... before it is so stunned or deprived of sensation.[74]

The proposed legislation made no mention of shechita. It was a general attack on slaughtering practices in the state. Rev. Rowley, the principal drafter of the bill, argued that the impetus for the SPCA's move was the general inhumanity

71 Dr. Francis H. Rowley, "An Indictment of the American Slaughter-House," *Our Dumb Animals*, March 1912, 145, 157.

72 "Hebrews Win Over S.P.C.A.: Slaughter Bill Not to be Reported to Legislature," *Boston Post* March 3, 1911; "Mass S.P.C.A. Tries to Pass Anti-Shechita Bill," *Jewish Advocate*, March 2, 1911; "Jews Oppose S.P.C.A. Bills," *Boston Post*, March 2, 1911.

73 "Humane Slaughtering," *Our Dumb Animals*, April 1911, 168.

74 Bill 738, Massachusetts, 1911, "An Act Relative to Slaughtering Animals" in "Mass. S.P.C.A. Tries to Pass Anti-Shechita Bill," *Jewish Advocate*, March 2, 1911.

of slaughtering practices in Massachusetts. "He realized, however, that the bill would affect the Jews for whom he professed the utmost respect, but he felt that the larger measure of humanitarian should outweigh any religious consideration."[75]

In his account of the hearing before the legislative committee and his pleas in favor of the bill, Rowley made it clear that much of the SPCA's argument was based in reports from the British Admiralty in 1904, and from humane societies in Switzerland and parts of Germany, where anti-Jewish motives were at the forefront of bans on slaughter without stunning. Citing a German study in particular, Rowley informed the elected representatives of Massachusetts that only one understanding of the Jewish method was possible. "The report concludes (a) that the present-day non-stunning method of slaughtering, as used by the Jews, should be strictly forbidden in a country of enlightened morality; and (b) that of all methods now necessary to secure meat none is so uniformly condemned by experts, and so distinctly characterized as reprehensible and out of harmony with the spirit of the times, as the non-stunning method."[76]

The discourse adopted by the SPCA leader is one in which humane considerations operate as the primary motivations for attempts to prohibit slaughter without stunning, and they are opposed to mere religious practices and beliefs. The SPCA's legislative proposal targeted both the actual practice of killing with the single cut of the shochet's blade, and the preparatory process of casting whereby the animal's leg was bound and it was placed in a position where the killing process could occur quickly and efficiently according to religious law. While Rowley was willing to accept that the proposed legislation would have a direct impact on the Jewish method of killing, he was a much stronger advocate for its prohibition on humanitarian grounds. The lack of religious freedom was a small price to pay for humanity. The Jewish communities of Massachusetts recognized that the bill could only be understood from their perspective as a direct and unmitigated attack on their basic beliefs and religious practices. "On Wednesday noon of this week practically not a Jew in Massachusetts knew that the Mass. S.P.C.A. had conceived the idea of getting the Legislature to adopt a measure which, did it become law, would have prohibited Shechita in this state."[77]

---

75 Ibid.

76 "Humane Slaughtering," *Our Dumb Animals*.

77 Ibid.; "The Shechita Question," March 10, 1911.

Rowley was aware of the direct effect of his proposed ban of slaughter without stunning on the Jewish community, but he chose to proceed without notice to the Jews of Boston. It was left to the editor of Boston's Jewish newspaper "by continuous use of telephone and telegraph" to inform Jews throughout the state and to raise communal resistance to the proposed measure. The proposed SPCA bill "was dangerous to the liberties of the Jews, and at the same time inferred that they were not humanitarian."[78] These years in the first two decades of the twentieth century marked the emergence of Irish Roman Catholic political power to challenge the traditional Yankee establishment of Massachusetts Protestantism, perhaps embodied by Rowley. The same time period witnessed an evolving and significant Jewish vote in Boston.[79] The mass turnout to the state legislature hearing room included members of the Jewish community, as well as the state representative from the Jewish section of Boston.[80] Spokesmen for the slaughterhouse industry intervened in opposition to the bill.[81] Despite the SPCA's best efforts, the new political powers of Massachusetts and Boston, together with vested industrial interests, were allied against slaughterhouse reforms.

While recognizing, publicly at least, the good faith and sincerity of Rowley and the SPCA in proposing the measures in the bill, a leading Boston rabbi, M. M. Eichler, who had come to that city from Philadelphia, attacked the measure both in scientific and theological terms. As had other witnesses, he pointed out that there was long-standing scientific evidence that shechita was at least as humane as any other method then in use. Should the bill come into effect, "it would be the most anti-Semitic measure that had ever been passed by any American legislature."[82]

After the representatives of those opposed to the bill had spoken and Rabbi Eichler had put paid both to the scientific substance of the argument advanced by the SPCA and to any claim that this was nothing more than legislation offered for the same purposes, a show of hands indicated that only four members of an audience of dozens, still supported the legislation. Rowley entered into belated negotiations with the Jewish community. He proposed both a formal exemption for the Jewish method as part of Section 1 dealing

78 Ibid.

79 Meaghan Dwyer-Ryan, "Ethnic Patriotism: Boston's Irish and Jewish Communities, 1880–1929," PhD dissertation, Boston College, 2010, 68–108.

80 Braverman, *The ascent of Boston's Jews,* 237–238.

81 "Anti-Shechita Bill in Massachusetts," *Jewish Exponent,* March 10, 1911.

82 "Mass. S.P.C.A. Tries to Pass Anti-Shechita Bill," *Jewish Advocate.*

with the requirement of stunning, and suggested and promised further discussions about the problems related to casting set out in Section 2. Because of the absence of any real statewide representative body for Massachusetts' Jews, the discussions took place largely between Rowley and the editor of the *Jewish Advocate.* The Jewish representative was reluctant to accept any idea of a Jewish exemption if that would imply that the Jewish method was cruel and inhumane, but was placed outside the legal prohibition on killing without stunning purely on religious accommodation grounds. Jewish communities always insisted that shechita was necessarily conducted in a way that showed respect for the animal to be killed and that the method involved little, if any, suffering.

The idea of creating better practices for casting, including investigating technological solutions to enable the mechanical placing of the animal in a position where it could be killed by the shochet, was seen to be unproblematic by the Jewish representative. Jewish authorities insisted that shechita and Jewish law and ethics always demanded the highest degree of respect for the animal to be killed, and a sincere desire to lessen suffering. Reducing pain or fright by improving casting was perfectly consistent with Jewish belief and practice.[83]

Rowley had been disingenuous in his proclaimed desire to reach an appropriate compromise. In the pages of *Our Dumb Animals* he continued to assail "Hebrew opposition" to humane slaughtering and the SPCA's attempts to make Massachusetts a more decent place to live. Not only did he attack the inherent cruelty of shechita, but he rejected the theological justifications of the Jewish method when he reasserted that there was no firm religious basis for the practice.[84] Rowley specifically addressed Rabbi Eichler's comments to the legislative committee and refuted any idea that the SPCA's proposal was motivated by antisemitic sentiment, in terms that could be considered antisemitic. "No one, outside the Jewish community, would dream that any legislation put forward in the interests of a nobler humanity was inspired by anything so bitter and unchristian as anti-Semitism."[85]

The assertion by a leading animal welfare advocate that their cause was inspired by a nobler humanity offers the familiar taxonomy in which humanity and now nobility can be opposed to the Jewish community, in a manner in which he states antisemitism is completely absent. This is perhaps

83 "Shechita Bill to Have Been Amended," *Jewish Advocate,* March 10, 1911; "The Shechita Question in Boston," *American Hebrew,* March 10, 1911.

84 "The Jews' Opposition," April 11, 1911, 169.

85 "Replying to the Rabbi," ibid.

a debate in which intention and effect are opposed in the minds of the two parties, but Rowley's baseline that shechita was inherently cruel was, in his mind, a simple scientific and human truth that Jewish assertions of religious freedom could never trump. When Rabbi Eichler, to the approval of the audience, called upon the traditions of the foundational American value of religious freedom, Rowley replied, in terms that repeated the absurd equivalence in the United Kingdom between the defense of shechita and the Hindu practice of suttee. He invoked more recent rejections of Mormon polygamy, claiming that such religious practices were banned because of the "outraged moral sense of their fellow-citizens." For Rowley, Jewish claims of religious tolerance were ill-disguised manifestations of the tyranny of the minority. "Would it not be far more unjustifiable tyranny for some 150,000 of our citizens to insist that their customs should prevail against the sentiments and desires of those who outnumber them twenty times?"[86]

The taxonomical structures of Rowley's narrative are at least consistent. The needs and desires of humanity cannot be stymied by the customs of a minority religion that seeks to invoke American values to hide the inherent cruelty of their foreign practices. Jews continued to insist on practices that, "in the judgment of multitudes of men and women of the finest sensibility and character (were) unnecessarily cruel and painful."[87] Humanity, sensibility, character, and civilization stood in stark opposition to cruel Jewish practices in the slaughtering of animals for meat. Jews were, as they had always been, outside any possible construction of humanity. Massachusetts Jews understood the real message behind Rowley's arguments and attacks on shechita.

> It is a strange coincidence that almost every so-called humanitarian who has so far assailed the Jewish mode of slaughter has been, demonstrably an anti-Semite in disguise. It would be wrong to accuse our present critic of such leanings, though men as broad as the historian Rhodes have, to their amazement, found themselves convicted of unconscious anti-Jewish prejudice.[88]

The universalizing teleological construction of cruelty articulated unremittingly by Rowley placed Jews in the category of the inhumane as long as they

---

86 Ibid.

87 Ibid.

88 "Cruel Mercy," *American Israelite*, November 14, 1912.

ignored scientific, theological, and social fact, and insisted on their ill-conceived continuation of custom disguised as religion. Fortunately for the Jewish communities of Massachusetts, the state legislators acted in a manner more consistent with a different understanding of American constitutional values and rejected Rowley's pleas.

The SPCA withdrew its proposed legislation.[89] After the failed imposition of a ban on shechita in the state, the Jewish communities in Massachusetts saw that organized opposition to measures that would have a negative impact on their rights and religious practices could be effective. The need for a more formal communal structure, a *kehilla,* was now more urgent than ever.[90] Immediately after the victory in preserving shechita, the Jews of Massachusetts formed a not completely successful state wide representative body to protect communal interests.[91] But the combat of the Massachusetts SPCA against the Jewish method was not over. Rev. Dr. Rowley mounted a nation-wide attack.[92]

The Jewish community had fought back, advocating the humane nature of the Jewish mode of killing animals for human consumption, as demonstrated by the latest scientific evidence.[93] Despite this ongoing opposition from the Massachusetts Jewish community, the SPCA under Rowley's leadership persisted in its claims against shechita.[94] Believing that the SPCA was about to mount yet another legislative assault on the Jewish method, the *Jewish Advocate* launched a fund-raising campaign to protect Jewish religious observance. The editor made it clear to readers that Massachusetts did not stand alone, and that the possible attempt to ban shechita had to be understood as part of an international campaign.

> The S.P.C.A. everywhere has the money, and it is making a world-wide attempt to suppress Shechita. The combat will take the form of presenting the official and expert statements, and, as it is not desirable that this struggle should be renewed annually, it should be conducted to a conclusion…

89 "Shechita Victorious," *Jewish Advocate,* March 31, 1911.

90 "Editorial: Keep Cool," *Jewish Advocate,* March 10, 1911.

91 "Jews Form a State Union," *Boston Post,* March 16, 1911; "Issac Heller Cries 'Hands Off' Kashruth," *Jewish Advocate,* August 2, 1912.

92 Herman H. Rubenovitz, "Shechita," *Jewish Advocate,* February 28, 1913; "Still Trying to Prohibit Shechita: Propaganda for the S.P.C.A. Makes Its Position Clear," August 2, 1912.

93 "The Humaneness of Shechita: No Cruelty in Jewish Method, Says Prof. Louis Ginzberg," *American Hebrew and Jewish Messenger,* October 1913; Baila R. Shargel, "Louis Ginzberg As Apologist," *Jewish History* 79, no. 2 (1989).

94 Max Heller, "Shechita Again," *American Israelite,* November 13, 1913

> The attack on Shechita is an attack on Judaism as a humane religion, whether the assailants design it so or not. To thousands of Jews the prohibition of Shechita would mean the prohibition of the whole dietary laws.[95]

Rowley and the SPCA did not introduce anti-shechita legislation in Massachusetts in the months following the propaganda campaign, but neither did they abandon the focus on shechita. The Society reported in joyful terms on the conviction of Rabbi Levitt in Halifax and the declared intention of the Nova Scotia SPC to use the case as a springboard to ban all slaughtering without stunning. Rowley wrote:

> We have hoped that our Jewish friends in this country would see the wisdom of yielding to modern humane sentiment and abandon their ancient ritual of slaughter which, outside the Jewish faith, is almost universally condemned as causing wholly avoidable suffering. If the plea is made that to prohibit it would be to interfere with religious liberty, one wonders if in the name of religion any citizen has a right to violate a law against cruelty. Certainly, that was not the court's opinion in Nova Scotia.[96]

Until it was, and a Nova Scotia court overturned the conviction, combining grounds of religious freedom and the preponderance of scientific evidence demonstrating that shechita was not inherently cruel. No mention of the ultimate outcome in Nova Scotia can be found in the pages of *Our Dumb Animals*.

Rowley, as Chair of the Slaughter House Reform Committee of the American Humane Society, some years later declared that he and his organization would be willing to offer a Jewish exemption for all killings conducted by shochetim.[97] The SPCA in Massachusetts had again introduced a bill, "Relative to the Slaughtering of Animals," echoing the English and Scottish legislative structures, and South Australian legislation, that included an exemption for killings carried out by "any member of the Jewish faith duly licensed by the Chief Rabbi as a slaughterer, when engaged in the slaughtering of cattle intended for the food of those of the Jewish faith according to the

95 "Will You Help?" *Jewish Advocate*, November 14, 1913; "Help Protect Shechita," November 21, 1913.

96 "A Notable Conviction," *Our Dumb Animals*, June 1913, 12.

97 Harry Schneiderman, "American Humane Societies Not Opposing Shechita," *American Israelite*, May 10, 1923.

Jewish method of slaughtering."[98] The *Jewish Advocate* pointed out that cutting and pasting from other jurisdictions caused at least one significant problem for the Massachusetts Society's proposal.[99] Unlike the United Kingdom, there was no Chief Rabbi in Massachusetts. The historical debates about the proper authorization of shochetim had arisen in London and elsewhere precisely because many rabbis, especially in conflicts between Ashkenazi and Sephardic traditions, had insisted upon their own inherent power to test and approve shochetim. While changes within the formal structures of British Orthodox Jewry, concretized in the powers of the Board of Shechita and the Chief Rabbinate, would alleviate these problems, these structures were not replicated in Massachusetts.

The SPCA in Massachusetts had consistently failed in its attempts to ban shechita, but it had not abandoned its unwavering opposition to the Jewish method. A more complete account of the legal history of anti-shechita agitation and prosecution in the Commonwealth of Massachusetts remains to be written, if and when archival records can be discovered. In the meantime, the narratives of Angell, Rowley, and the Massachusetts SPCA juxtaposing humanity and Jewish religious practice fit into the broader Anglo-American legal history of attempts to attack and prohibit shechita.

---

98 M.M. Eichler, "A New Shechita Bill in Massachusetts," *Jewish Advocate*, January 1, 1920.

99 "A Fallacy in Shechita Bill," *Jewish Advocate*, January 8, 1920.

CHAPTER 9

The End (or Not) of the Story: Shechita and the Law in the Anglo-American World, 1855–1913

The limited timeframe of anti-shechita prosecutions should not suggest that anti-shechita agitation in the Anglo-American world began and ended within this period. Anti-cruelty groups turned their attention to outlawing the Jewish method in legislation banning slaughter without stunning. The year after the Halifax case, a concerted effort to arouse public feeling against shechita and to garner support for compulsory stunning legislation was orchestrated by the RSPCA throughout England.[1] American organizations continued with sporadic attempts to introduce compulsory stunning for some time. Similar campaigns continue to this day. European antisemitism has always targeted the religious method of slaughter, and continues to do so.

The Anglo-American story of anti-Judaism is not too far removed from continental counterparts as far as the project of banning shechita is concerned. No ban on the Jewish mode of slaughter has been put into effect in Australia, Canada, the United Kingdom, or the United States, either through litigation, or by legislative fiat. This does not mean that animal welfare groups, led historically by the RSPCA and its related organizations, have ever abandoned their belief that the compulsory introduction of pre-slaughter stunning or killing, by bolt or electric shock, or in an early exploration, gassing the animals, is a central goal of humane slaughter.[2] Despite Jewish theological assertions about holy commands to abjure cruelty, and clear historical evidence of Jewish

1 "Shechita," *JC*, June 26, 1914 "Attacking Shechita," May 29, 1914, "Shechita Attacks," May 29, 1914; "Attacking Shechita," May 1, 1914.

2 Henry MacCormac, *The Painless Extinction of Life in Animals Designed for Human Food* (London, 1864).

is a juridical embodiment of core values of religious toleration, acceptance, or even of religious liberty. The exemption offers a protection for what is considered as an essential element of the Jewish religion by many Jews. The provision in question carves out a juridical space for the free exercise of religion, and for Jewish self-regulation in the practice of shechita.

A second reading of the legal meaning and function of the exemption for the Jewish mode of slaughter both recognizes and restricts this space of religious freedom. The argument harks back to discussions about the state of scientific knowledge concerning animal pain and suffering, and the cruel infliction of unnecessary pain banned by animal protection legislation. The exemption is a legislative recognition of part of the argument put forward consistently by the Jewish community on the shechita question. The assertion that the Jewish method is in the alternative more humane than, or as humane as, any other method for killing animals for human consumption, is concretized in the legislative provisions in many jurisdictions exempting shechita from generalized stunning or instantaneous killing requirements. The problem with this position is readily recognizable and is given voice within the standard legal understandings of religious freedom in the English legal system. Poulter exemplifies the traditional view:

> Their (Jews and Muslims) traditional concern for animal welfare should not be belittled and their freedom to practice their religions under international human rights conventions should be respected. If the scientific evidence were eventually to become overwhelming in its conclusiveness that such methods of slaughter involved unnecessary suffering to animals then Parliament could justifiably restrict religious freedom on the ground that the practice needed to be curbed in the interest of the protection of public morality....[19]

This idea, which in no substantively important way changes under the European Convention of Human Rights, or the Human Rights Act, or the Canadian Charter of Rights and Freedoms, or American First Amendment jurisprudence, embodies all the normative conflicts that have always informed debates about shechita. They have been repeated more recently and in

---

19 Sebastian Poulter, *English Law and Ethnic Minority Customs* (London: Butterworths, 1986), 281; Ari Mermelstein, Victoria Saker Woeste, Ethan Zadoff, and Marc Galanter, *Jews and the Law* (New Orleans: Quid Pro Books, 2014).

minority populations and their practices. The reasons for this are complex, multifaceted, and not by definition inevitable.[38] The history of the anti-shechita context offers an example of how, in different geographical and historical contexts, the slippage into antisemitic stereotypes and discourses manifested itself in unequivocal ways. The invocation of ethical boundaries and discourses of cruelty and humanity, particularly in the criminalizing context of prosecutions, led to the deployment and normalization of antisemitic tropes as part of legal practice. Law is politics, and anti-shechita legal practice embodies and has always embodied and given rise to an expanded politics in which anti-Jewish motives, *motifs,* themes, and consequences are played out within the context of a judicial discourse meant to insure the humane treatment of animals.

As Kymlicka and Donaldson assert, there is nothing inevitable about this appropriation of animal rights discourse in the philosophical sense, but there has been an historical and cultural determination to use the opportunities presented to engage in direct attacks on Jewish religious practice. Cruelty and humanity are juxtaposed and invoked as synonyms for Jewish and non-Jewish/Christian as a pernicious hatred becomes part of everyday legal practice. In the end, this is perhaps more than a mere unfortunate coincidence, and something closer to a David Rome's understanding of anti-shechita prosecutions as a "major attack on Jewish freedoms."

38 Will Kymlicka and Sue Donaldson, "Animal Rights, Multiculturalism, and the Left," *Journal of Social Philosophy* 45, no.1 (2014).

activism in the world of animal welfare, in England, Canada, Australia, and the United States, the attack on shechita has always imposed the label of innate cruelty on the Jewish method of slaughter. In the United States context, one that can be easily transposed and applied throughout the English-speaking world, Davis argues that, "by insisting on unconsciousness as the singular, incontrovertible litmus test for a merciful death, American humane leaders and their allies abroad tacitly denied the humane precepts of kosher slaughter."[3]

As a necessary consequence of this continuous position in relation to the core function of stunning in reducing animal suffering in slaughtering for food, SPCAs, Humane Societies, and other animal welfare organizations, have led unrelenting campaigns against shechita, and those attempts to ban the Jewish method of slaughter have not ended.

The narrative of the prosecutions of shochetim under general anti-cruelty provisions allows us to focus on the privatization of the enforcement of anti-cruelty legislation in the hands of SPCAs. There is another aspect of the privatization issue in relation to shechita that is the necessary, if occasionally neglected, concomitant of the penal role of the private animal welfare organizations. That obverse element is the way in which Jewish communities, even as they have come under stress when faced with attempts to criminalize shechita, have maintained control over the practice. This is not to suggest that the history of the Jewish method of slaughter is an unambiguous tale of Jewish self-governance. The record of the so-called kosher libel cases demonstrates in a number of jurisdictions the struggles within and between Jewish communities for an effective and accepted form of internal self-regulation of the kashrut process.

In sections of the non-Jewish community, for all its difficulties and controversies, the general process of kosher certification, of which shechita is an important element, has come to embody a focus on hygiene at the heart of Jewish law relating to the preparation of food.[4] Kosher certification has become for many modern consumers a guarantee of food safety and hygiene.[5] The history and debates that have surrounded kosher certification in the United States in particular, with significant failures in terms of labeling and food safety, and even with ongoing conflicts, tell another story of private regulation of

---

3 *The Gospel of Kindness*, 98.

4 Adler, "Sanitation as Taught by the Mosaic Law."

5 Courtney I.P. Thomas, *In Food We Trust: The Politics of Purity in American Food Regulation* (Lincoln: University of Nebraska Press, 2014).

matters of public interest, and mark out kashrut as an important and complex American success story in private/public regulatory practice.[6]

Intra-Jewish organizational battles and disagreements do not hide the fact that the practice of shechita has been, and continues to be, a central element of Jewish communal self-identity, even for those Jews who do not keep kosher.[7] Attacks on the Jewish method of slaughter have always been, and are, perceived and understood to be assaults on a broad idea of Jewish identity in which kashrut and shechita are key signifiers of Jewish belonging.[8] Even for non-observant Jews, these questions and attempts to limit access to kosher food through limitations or bans on the Jewish mode of slaughter, are experienced as specifically targeting aspects of their identity that are Jewish.[9] The fundamental values of ritual, purity and impurity, and ethical demands relating to life and death, form part of the signifying chain that maintains core textual and social meanings within a general semiotics of Jewishness that transcends actual religious practice or belief, and moves to a level of historical, communal identity, and identification.[10]

The stories of animal cruelty legislation and the creation of a privatized enforcement mechanism as a central element of Societies for the Prevention of Cruelty, and attempts to prosecute shochetim embody many important political, social, and legal elements, in terms of the core values of religious freedom and the protection of shechita. These cases are exemplars of the nature, form, and function of legislation, the state, and the public/private distinction in Victorian and Edwardian times, or in America, in the Gilded Age.

Anti-cruelty legislation displays the arrangements and compromises made in light of an emergent state apparatus, in which functions from policing to hygiene, sanitation, and animal welfare emerged as concerns of the

---

6 Horowitz, *Kosher USA*; Timothy D. Lytton, *Kosher: Private Regulation in the Age of Industrial Food* (Cambridge, MA: Harvard University Press, 2013).

7 David C. Kraemer, *Jewish Eating and Identity Through the Ages* (New York: Routledge, 2007).

8 Stanley Waterman, "Eating, Drinking and Maintenance of Community: Jewish Dietary Laws and Their Effects on Separateness," in *The Changing World Religion Map*, ed. S.D. Brunn (Dordrecht: Springer, 2015).

9 Florence Bergeaud-Blackler, Ari Z. Zivotofsky, and Mara Miele, "Knowledge and Attitudes of European Kosher Consumers as Revealed through Focus Groups," *Society and Animals* 21, no. 5 (2013); A. Memmi, W. Ackermann, and S. Zoberman, "Pratique Religieuse et Identité Juive," *Revue française de sociologie* 14 (1973).

10 Sophie Nizard-Benchimol, "L'abattage dans la tradition juive," Études rurales 147–148 (1998).

government, while at the same time being delegated to private animal welfare groups in the case of anti-cruelty regimes. This delegation emerged both because in the earliest stages, the state bodies were newly formed, organizationally and ideologically weak, and because single-issue welfare organizations, such as the SPCAs, had the ability and desire to focus on enforcement within their narrow remit. In its earliest days, the organization sometimes found itself in conflict with judicial authorities. There was more often than not a need for corroborative testimony from sworn police officers to convince magistrates of the case before them. "On the whole, however, relations between the animal protection movement and the law enforcement establishment were good. Indeed, the RSPCA was unusual among nineteenth-century reform movements in maintaining a cooperative rather than an antagonistic relationship with officially constituted authorities."[11]

A more detailed socio-legal historical study of anti-cruelty laws and the institutional enforcement arrangements putting them into practice, with the creation of an army of inspectors and prosecutors dedicated to a single, legal, moral cause could offer important insights into the nature and shape of the public/private distinction in the era of massive social dislocation, economic change, and new, and sometimes innovative concepts of welfare, and mechanisms of governance, many of which persist today.

This historical embodiment of a particular set of institutional arrangements serves as a key example for those concerned with the juridical and economic nexus of modern legal systems. Lawrence has argued that while the mustering of political will through successful lobbying for the introduction of anti-cruelty legislation was a singular success of the SPCA in the United States, only through private enforcement powers could the legislation have been put into effect. Political will to make a symbolic statement about the core values of anti-cruelty is one thing, but allocating scarce governmental resources to enforcement of the normative vision is an entirely separate matter. "If humane society agents had not initially been delegated the authority to enforce animal cruelty statutes, those statutes may well have gone unenforced. The political pressure sufficient to enact the statutes was probably inadequate to raise the funds to enforce them."[12]

---

11 Harriet Ritvo, *The Animal Estate: The English and Other Creatures in the Victorian Age* (London: Penguin, 1990).

12 David M. Lawrence, *Private Exercise of Governmental Power* 61 Indiana Law Journal 647 (1986), 657.

That type of enforcement through a private agency such as SPCAs grants extensive powers to a body that is not part of the state organizational structure in the sense of being a public body. It is a private organization exercising public functions of policing animal welfare both in the private sector i.e. by individuals, and now in parts of the public sector, as the RSPCA supervises the welfare of police dogs.[13] Scott accepts the RSPCA regulatory function and activities as an example of the mix of standard setting by public authorities, and enforcement by a private body. "For humanity, the private monitoring and enforcement of public rules against cruelty to animals provides an example."[14]

Scott's assertion needs fuller contextualization and more careful analyses grounded in and informed by the history of the anti-shechita prosecutions. One key problem that has emerged throughout the cases in which SPCAs sought to put an end to the Jewish method of slaughter has been the vagueness of the concepts and categories of behavior punishable under the different statutes. General terms such as cruelty and ill-treatment, qualified as unnecessary or wanton, combine to make the actus reus and mens rea of the alleged offenses committed by the shochetim extremely difficult to prosecute and to defend. While the vagueness of the terminology, and traditional criminal law values of the burden of proof, have generally served the Jewish officials accused of animal cruelty, they have informed the contexts of the cases and permitted the SPCAs to bring the prosecutions in the first place. Uncertainty in the hands of the zealot with a univocal moral agenda can be, and has been, a dangerous thing.

Such general terms combined with a single-minded raison d'être inform the functions and goals of SPCAs. Their historically continuous obsession with concepts of humane slaughter, and the never-ending struggle against what is for them the barbarity of shechita, with its theologically driven refusal to contemplate stunning, laid the groundwork in which the prosecutions, or persecutions, of shochetim became inevitable. Scott believes that the RSPCA exercises its public functions for humanity, and this may well be true in many instances of its enforcement practices aimed at stamping out cruelty to animals. The evidence clearly reveals that the single-minded, narrow ideological focus of the RSPCA and its fellow organizations in other jurisdictions, has led to campaigns with clear antisemitic effect against the practice of shechita.[15]

13 Colin Scott, *Private Regulation of the Public Sector: A Neglected Facet of Contemporary Governance*, 29 Journal of Law and Society 56 (2002), 62–63.

14 Ibid., 69.

15 Ian MacLachlan, "*Coup de Grâce:* Humane Cattle Slaughter in Nineteenth Century Britain," *Food and History* 3, no. 2 (2005), 166–167.

The problem with the privatization of such a function is that when it is informed by a normative and morally infused mission to protect God's helpless creatures, it privatizes the essential public function of determining and enforcing a different, social, and political ethos of humanity, of a demos, of a body politic, a key part of which in liberal democracies has been the respect for, and protection of, core religious beliefs and practices of citizens. The needs of humanity, even in so far as the treatment of animals is concerned, may simply be too complex, and too politically, socially, theologically, and legally important to be delegated to those on a mission from an overwhelmingly Protestant God. The protection of humanity, or acting as humanity's avenging agent, can never safely be placed in the hands of an organization with such a narrow vision and understanding of the transcendent ethical norms involved in the practice of shechita.

Agitation around kosher meat (and increasingly around halal practices) under the guise of humanity, occurs in political, social, legal, and cultural contexts of increased xenophobia, whether around the Aliens Act at the turn of the twentieth century, or the migrant crises of today, anti-Zionism/antisemitism, appeals to core Englishness or Britishness, or American/Australian/Canadian values. In the presence of weak governmental and civil society responses to the undermining from left and right of ideas and ideals of solidarity and equality, attacks on religious beliefs and practices impose important ethical demands on citizens of democracies. The undercurrent of antisemitism in anti-shechita programs and policies has been a central component of these socio-legal histories from the beginning and throughout the Gilded Age and beyond.

Historically, the Jewish response to claims of animal cruelty has always been couched in a way that offered a direct reply to these claims.[16] Relying on centuries of experience by shochetim, and testimony from veterinarians, and subsequent scientific investigations into the physiology of pain, litigation and political strategies have converged at the point where advocates for maintaining Jewish practice have sought to prove that shechita is not cruel. They have presented evidence that death is quick and as painless as possible when placed in the hands of a skilled shochet wielding a *hallaf* sanctioned by Jewish law. In some cases, they managed to convince courts and legislatures that there was scientific evidence to support the argument of the absence of cruelty in the Jewish method, often by pointing to the barbarity of clubbing, or pithing, of failed attempts using electricity or bolts, or bullets. This discourse of

16 Ronald Kaye, "The Politics of Religious Slaughter: Strategies for Ethno-religious Political Action," *New Community* 19, no. 2 (1993).

non-cruelty was often accompanied by the slightly different comparative position of no more cruelty than any other method. This became more and more jurisprudentially and politically difficult as the RSPCA and its fellows developed better technological methods of humane slaughter. They insisted over and over, through the years, within the prosecutions, and underpinning the attempts to institute legislative change, that stunning before killing was the only way to remove cruelty and ill-treatment from slaughtering practices. The question of a Jewish exemption then reared its head, placing shechita in an awkward, and inescapable position of juridical otherness.

The instantiation of this legalized Jewish otherness can be seen in the first national slaughtering legislation in England, the 1933 Slaughter of Animals Act, described as providing "for the humane and scientific slaughter of animals."[17] Some five years earlier Scottish legislation on the same subject had made special provisions protecting "the Jewish method of slaughter."[18] The English statute, like its Scottish counterpart, imposed stunning but made special provision for animals slaughtered "by the Jewish method for the food of the Jews and by a Jew duly licensed for the purpose by the Rabbinical Commission…." (§1(b)(i). But the Jewish exception was not blanket permission since it still imposed a statutory ideal of an animal "slaughtered without the infliction of unnecessary suffering" by the Jewish method. (§1(b)). The idea embodied in the statutory language, its underpinning hermeneutic framing, is clearly that the Jewish method might still result in unnecessary suffering and only those instances of shechita where this was not the case would be permitted. Instantaneous killing or stunning did not suffer from any such qualification.

While there is still some ambiguity in the legislative language, a more significant doubt informs the ideological socio-legal environment of the permission or toleration granted to shechita. The Jewish method is legislatively framed as other than the usual methods of stunning or instantaneous killing, which are, by definition humane and scientific. There is no explicit, statutorily embodied, and confirmed recognition of the historic arguments about the humanity and non-cruel nature of shechita. Instead the practice is legislatively marginalized. It is permitted and tolerated, but it can never be juridically normalized. It remains ineffably Jewish. A somewhat more positive reading of the specific inclusion of a Jewish exception might take one of two interpretive turns. One might argue that the inclusion of a Jewish (and a Muslim) exception

17 23 & 24 Geo. V, c. 39.

18 Slaughter of Animals (Scotland) Act 1928, 18 & 19 Geo. V, c. 29, §§ 2 (6), 8.

This leads us to a legal historical and political, cultural, and social irony. Herman's study of consistent Jewish failure to advance successful equality claims examines cases that largely occurred after Jewish emancipation. The anti-shechita cases occurred at an earlier time when emancipation and equality for Jewish immigrant communities were much more open-ended and undecided concepts. As Jews have become more formally and substantively equal citizens, it has proved increasingly difficult, if not impossible, for them to succeed in establishing illegal discrimination.

To Herman's critical analysis of English law, it is necessary to add these anti-shechita cases and to see them as juridical embodiments in Australia, Canada, Scotland, and the United States as examples of the same difficult taxonomical and ethical struggles to find a place for Jews, and Jewish religious practice, among the legal normativities and normalities of these countries. A nuanced socio-legal history of "the Jew" within each of the legal systems in which anti-shechita prosecutions occurred, along the lines set out within Herman's analytic, is no doubt called for and perhaps urgently necessary.

When the institutional governance structures of law enforcement are privatized and the normative and legal frame of cruelty informs attempts to criminalize the Jewish method of slaughter, broader social, political, and jurisprudential understandings of the public sphere and our collective political values and the place of Jewish (and other) religious practice therein, and of the private sphere and ideas of religious autonomy are forgotten. When legal arguments are asserted, and regulatory theory unproblematically constructs a category– humanity– within which Jews are cruel and from which Jews and Jewish religious practice are excluded, the urgency of more fully and carefully articulated socio-legal histories becomes clear.

In a different but related context, that of the live chicken market in today's San Francisco Chinatown, Kim's analysis serves as a contemporary example for our socio-legal understandings of the historical anti-shechita cases. Describing opposition to the sale of live animals, she concludes that "while their arguments for the most part scrupulously respected multiculturalist and colorblind norms, the optic of cruelty inevitability mobilized ethnocentric and anti-Chinese feeling among some part of the public."[37]

This can be expanded to highlight the ways in which many anti-cruelty causes were used to invoke majority racial constructions of ethnic and religious

37 Claire Jean Kim, *Dangerous Crossings: Race, Species, and Nature in a Multicultural Age* (Cambridge: Cambridge University Press, 2015), 100.

somewhat more detail by Renteln, but they already exist in a presupposed, but almost entirely unarticulated, taxonomical structure of Jewish otherness and cultural exoticism within legal culture.[20] They give voice to the continuing insistence by the RSPCA, among others, that humane slaughtering must involve instantaneous death or stunning before slaughter. There is a prima facie recognition of the values of religious freedom and tolerance. There is even a rejection of any antisemitic informing narrative in the acknowledgement by Poulter of the Jewish value of kindness and abhorrence of animal cruelty. Tolerance in this guise is an operation in otherness and in othering. It exists for Poulter only until such time as science and modern forms of knowledge, which are somehow made synonymous with Englishness and core national values, can demonstrate the reality of unnecessary suffering. An obviously Christian or humanist, but never Jewish, public morality takes over and a ban on shechita would be the duty of Parliament.

There is an intriguing deployment of the signifier unnecessary, one that clearly privileges science and knowledge at the expense of religious practice and belief. The suffering is only comparatively unnecessary because of technological advances in killing techniques. A moral position is proposed in absolute terms, end suffering, but it can only ever be portrayed in a relative sense—unnecessary. There is no room in the calculation of what is or is not necessary in terms of a possibly competing value i.e. religious practice and belief. The law as interpreted by the courts in relation to slaughtering for human consumption has always recognized that at some point, cruelty and suffering, however brief or transient, are inherent in killing sentient creatures for food. The position on the morality of killing animals must be contingent and subject to changes in technological capacity in relation to killing, from the poleax and club, to the bolt gun, to electrocution.

The argument defending shechita has always been posed in terms of the absence of unnecessary suffering.[21] This arises out of normative necessity relating to the taxonomies of legislative language and available legal argument, and a further political demand to counter social and theological discourses underpinning anti-shechita movements about Jewish blood lust and primitiveness. No room has been discursively available for the position implicit in some interpretations of the Jewish exemption provisions that there may be an

---

20 Alison Dundes Renteln, *The Cultural Defense* (Oxford: Oxford University Press, 2004), 101–102.

21 Graham Zellick, "Freedom of Religion and the Jewish Community in the United Kingdom," *Patterns of Prejudice* 21, no. 2 (1987): 14.

element of cruelty in shechita, as there is in all killing, but that such cruelty is outweighed in any normative calculus in a liberal democracy by a more important social value of religious freedom. This is not to suggest that shechita is cruel, or crueler than any form of killing animals for food, and is an idea put forward in full recognition that there are good reasons it has not been at the center of debates on the Jewish method. But it is a factor of some normative weight that cannot be ignored when confronted with assertions, from the earliest days of RSPCA activism, to Poulter's jurisprudential argument, that somehow there is a position of public morality and then there is the Jewish position.

The problem that arises whenever there is a Jewish exemption (or a Jewish and Muslim exemption) is that the religious method of slaughter is positioned as abnormal, extraordinary, and always precarious. It sits awkwardly, and as Poulter makes clear, contingently, in a difficult and transient relation between a formal legality, and a politically constructed, socially, and normatively doubtful legitimacy, an exception to the accepted, normal, standard, ordinary, scientifically infused methods (historically Gentile and/or Christian), a core value of American, Australian, Canadian, English, or Scots, identity, of humane slaughter.[22] In each of these countries, the tensions between religious freedom, into which one might place shechita, at least contingently, and the application of broad, general provisions, into which category we might fit anti-cruelty laws, has been and is the subject of evolving and, like the categories themselves, always contingent, jurisprudential debate. Two examples suffice to concretize the concerns.

The Canadian Charter of Rights and Freedoms establishes that everyone has the right to freedom of conscience and religion. (§ 2)[23] However, these rights are constrained by such "reasonable limits prescribed by law as can be demonstrably justified in a free and democratic society." (§ 1) Canadian courts have recognized extended freedoms in relation to Jewish religious practice, accompanied by a self-declared reluctance to inquire into individual motivation or belief, and a jurisprudential hesitation to engage in theological inquiries concerning belief and practice.[24] This jurisprudence results in what Shauna Van Praagh aptly characterizes as "a complex and continuing conversation

22 Noëlie Vialles, "Légitimités et Légalités: À Propos des Viandes Rituelles," in *A Croire et à manger: Religions et Alimentation*, eds. Sophie Nizard, Séverine Mathieu, Aïda Kanafani-Zahar (Paris: L'Harmattan, 2008).

23 *Canadian Charter of Rights and Freedoms*, Part 1 of the *Constitution Act, 1982*, being Schedule B to the *Canada Act 1982* UK, c.11.

24 *Syndicat Northcrest v Amselem*, 2004 SCC 47, [2004] 2 SCR 551.

about shared norms and understandings of coexistence."[25] As Benjamin Berger points out, these conversations involve complex sociological, theological, and legal practices that demand careful and well-articulated understandings about the meanings of the religious practices in question at any particular moment, and of an equally deep comprehension of any justification of legal limits on those practices as might be necessary in a free and democratic society.[26]

There is nothing in Canadian constitutional jurisprudence that would per se prevent the presentation of an argument that general public norms relating to preventing cruelty to animals should trump any claim that Jewish religious practice be exempt from their application. The understanding and conversation elements of Canadian law on freedom of religion do not preclude the creation of an argument that Jewish religious practice in relation to shechita is cruel and that a limit, or prohibition on that practice, is justifiable in a free and democratic society. This is and has always been the fundamental assertion of SPCAs on the question. General values of humanity must trump any special pleading for the right to violate laws that prohibit cruelty to animals during slaughter. These claims have led to the series of counter-arguments presented by Jewish authorities throughout the history of anti-shechita prosecutions/persecutions directly addressing and refuting assertions that the Jewish method is cruel.

Such arguments can only serve to illustrate the fact that as a religious, social, and political minority, Jews always find themselves in a position of contingent jeopardy. In today's Canadian constitutional discourse, demands for recognition of shechita are commonly constructed in terms of reasonable accommodation made by the majority to minority religious practice.[27] This is because at an important level of social hermeneutics there is an understanding of the abnormal character of the Jewish method of slaughter. As Ira Robinson has pointed out, in significant parts of the Canadian constitutional, political, and legal experience, ideas of reasonable accommodation themselves have been constructed in ways that manifest a positioning of Jewish otherness that is deeply ahistorical and reflective of a dominant ideology in which antisemitism has often been

25 "View from the *Succah:* Religion and Neighbourly Relations," in *Law and Religious Pluralism in Canada*, ed. Richard Moon (Vancouver and Toronto: University of British Columbia Press, 2008), 38.

26 *Law's Religion: Religious Difference and the Claims of Constitutionalism* (Toronto: University of Toronto Press, 2015).

27 Jean-François Gaudreault-DesBiens, ed. *Le Droit, La Religion et le "Raisonnable"* (Montreal: Thémis, 2009).

embedded.[28] Challenges to the dominant order, of which shechita has been, and could again be, constructed as a key example, by those who claim to speak in the name of humanity could easily, in the right political and social environment, come to "encounter the outer limits of law's accommodation, and it is a cold place to reside outside of the new wall of separation."[29]

Nor do the First Amendment protections of freedom of religion ineffably lead to a different result for American Jews. The Supreme Court decision in *Employment Division v. Smith* confirmed that statutes of general application could constitutionally interfere with religious practice as long as that interference appears to be an incidental effect of the legislation.[30] In a case involving a blanket ban on slaughter without stunning, passed by the legislature with the goal of ensuring the humane treatment of animals in the process of killing them for human consumption, the argument would be between those who assert that the principal and only real goal of such a law would be to enshrine broad understandings of a humanitarian morality, and those who would contend that the real and immediate effect of the legislation would be to directly prevent observant Jews from practicing their religion.[31] Direct targeting of a religious practice, even under the guise of general laws, is unconstitutional.[32] But the line between direct and indirect effects is not a bright one, nor is the one between religious belief and religious practice which historically have been treated differently, with the former enjoying much wider protections. This is not to suggest that shechita is in imminent danger or that the limited and limiting *Smith* test remains unchallenged or unchanged by other decisions.[33] But

28 "The Bouchard-Taylor Commission and the Jewish Community of Québec in Historical Perspective," in *Religion, Culture, and the State*, eds. Howard Adelman and Pierre Anctil (Toronto: University of Toronto Press, 2011).

29 Ran Hirschl and Ayelet Shachar, "The Constitutional Boundaries of Religious Accommodation," in *Constitutional Secularism in an Age of Religious Revival*, eds. Susanna Mancini and Michel Rosenfeld (Oxford: Oxford University Press, 2014), 191.

30 494 U.S. 872 (1990); Christopher L. Eisgruber and Lawrence G. Sager, *The Vulnerability of Conscience: The Constitutional Basis for Protecting Religious Conduct*, 61 University of Chicago Law Review 1245 (1994); Michael W. McConnell, *Free Exercise Revisionism and the Smith Decision*, 57 University of Chicago Law Review. 1109 (1990).

31 Anna Joseph, "Going Dutch: A Model for Reconciling Animal Slaughter Reform with the Religious Freedom Restoration Act," *Journal of Animal Ethics* 6, no. 2 (2016).

32 Church of the Lukumi Babalu Aye, Inc. v. City of Hialeah, 505 U.S. 520 (1993).

33 Burwell v. Hobby Lobby Stores Inc., 134 S. Ct. 2751 (2014); Hosanna-Tabor Evangelical Lutheran Church v. EEOC, 131 S. Ct. 1783 (2011); Caroline Mala Corbin, *The Irony of Hosanna-Tabor Evangelical Lutheran Church and School v. EEOC*, 106 Northwestern University Law Review Colloquy 96 (2011).

it is meant to highlight the more general idea that Jewish religious practice has been consistently defined within broad jurisprudential practice and discourse, as it has been by anti-animal cruelty groups, as an exception to the Christian rule of law.

Arguments such as those advanced by Poulter, the RSPCA, prosecutors in London, Manchester, Sydney, Aberdeen, Birkenhead, Denver, Attleboro, Atlantic City, and Halifax, are those put forward by more blatantly antisemitic agitators elsewhere.[34] These positions can only be asserted because of a core and underlying flaw in the legal system that in its operation creates and perpetuates the otherness of Jews, their beliefs, and practices, thereby creating the indefeasible dichotomy of public morality and a somehow opposable Jewish morality. Poulter's work is entitled *English Law and Ethnic Minority Customs*. The title embodies crucial dichotomies—English law and ethnic minority customs; English and ethnic minority being the two most significant and obvious. If we ignore the problematic confusion inherent in the sloppy deployment of the signifier ethnic in relation to religious practice, and all that is missed and missing in such a crude semiotics, we come to the core problem of the entire history of attempts throughout the English-speaking world to criminalize shechita by using prevention of cruelty legislation.

The Jewish method of slaughter is always constructed as other. There is the accepted way of killing animals for human consumption, the normal way, the Gentile way, the Christian way, the way of humane slaughter– poleax, club, gassing, pithing, revolver, electrocution, bolt gun– and then there is the Jewish mode, always, like the Jews themselves, other. Shechita is an ethnic minority custom and one that fits uncomfortably within English and other legal systems, or even evolving constructions of human rights.[35] The Jews, despite their long presence, their formal and to a lesser extent perhaps, their social emancipation, are for some never English (or Scots, or Canadian, Australian, or American) in an essential way. The cases demonstrate a continuing struggle for Jewish religious freedom in the face of a public morality, a humanity, a Christianity-infused campaign, a crusade, to impose a set of overarching moral values on what is constructed as a crude, primitive, and cruel people and their barbaric, bloody, religious practices. It is little wonder that Jewish communities have experienced these juridical instances as persecutions, not prosecutions.

---

34 Brian Klug, "Ritual Murmur: The Undercurrent of Protest Against Religious Slaughter of Animals in Britain in the 1980s," *Patterns of Prejudice* 23, no. 2 (1989).

35 Sebastian Poulter, "Ethnic Minority Customs, English Law and Human Rights," 36 *International and Comparative Law Quarterly* 589 (1987).

The limited history of debates in the courts over the alleged cruel and inhumane treatment of animals in the practice of shechita embodies the struggles of Jews throughout the Anglo-American world to combat a legal system and its ideological and normative structures that refuse to grasp the fundamental socio-political and juridical stakes.

> The question of the relationship between race, ethnicity, nationhood, peoplehood, and religion in English law (and many other questions) cannot be resolved, or even properly addressed, without recognizing not just the history of Jewish people and Jewishness in English law, but also the contemporary dominance of Christianity in thinking about English values and national character. In the meantime, Jewish and other people in England will continue to live in the shadow of Christianity's enduring force—yet another 'unfortunate coincidence'.[36]

Herman's study establishes the ways in which Jews inevitably lose when asserting claims under today's equality and human rights norms before English courts. The anti-shechita prosecutions are more complex. Joseph Myers and Abraham Friedner were convicted at Birkenhead. Others, like Yankoff Cohen and Philip Philipstein were acquitted; James Littman faced the bizarre Scottish verdict of "not proven"; Wolf Katznelson was convicted of animal cruelty in relation to shechita before the Massachusetts courts, as was Morris Kaplowitz in Atlantic City. A combination of factors appears to have influenced the acquittals in particular. Emerging discourses of emancipation for new Jewish immigrants probably played an important role in the not guilty verdicts, as religious freedom and doctrines of liberal equality emerged and took hold. A certain liberal Christian acceptance of appeals to millennial religious practice, of holy command, no doubt existed as part of a philosemitic normativity abroad among educated legal and political elites in London in 1855. One might consider the different purely legal context of the anti-shechita prosecutions. The prosecution carries an important and vital burden of proof beyond a reasonable doubt. The technical statutory language of anti-cruelty legislation, when combined with standard rules of burden of proof, may have acted to the benefit of the Jewish accused in many cases.

36 Didi Herman, *An Unfortunate Coincidence: Jews, Jewishness, and English Law* (Oxford: Oxford University Press, 2011), 174–175.

Bibliography

Abrams, Nathan. *Caledonian Jews: A Study of Seven Small Communities in Scotland*. Jefferson, NC: McFarland, 2009.

———. "Jute, Journalism, Jam and Jews: The Anomalous Survival of the Dundee Hebrew Congregation." *Northern Scotland* 3 (2012): 86–97.

Adelman, Howard, and Pierre Anctil, eds. *Religion, Culture, and the State*. Toronto: University of Toronto Press, 2011.

Adler, Hermann. "Sanitation as Taught by the Mosaic Law." *British Medical Journal* 2 (November 11, 1893): 1069.

Alderman, Geoffrey. "The British Chief Rabbinate: A Most Peculiar Practice." *European Judaism* 23, no. 2 (1990): 245–258.

———. "English Jews or Jews of the English Persuasion? Reflections on the Emancipation of Anglo-Jewry." In *Paths of Emancipation: Jews, States, and Citizenship*, edited by Pierre Birnbaum and Ira Katznelson, 128–156. Princeton: Princeton University Press, 1995.

———. *London Jewry and London Politics, 1889–1986*. London: Routledge, 1989.

Allen, Matthew. "The Politics of the Pew: Faith, Liberty, and Authority in a Sydney Church in 1828." *Journal of Religious History* (2016): doi: 10.1111/1467-9809.12407.

Amato, Sarah. *Beastly Possessions: Animals in Victorian Consumer Culture*. Toronto: University of Toronto Press, 2015.

Anderson, Jerry L. "The Origins and Efficacy of Private Enforcement of Animal Cruelty Law in Britain." *Drake Journal of Agricultural Law* 17, no. 2 (2012): 263–310.

Angell, George T. *Autobiographical Sketches and Personal Recollections*. Boston, 1893.

———. *Protection of Animals*. Boston, 1887.

Anonymous. "Anaesthetisation of Cattle Before Slaughter." *British Medical Journal* 1, May 31, 1884, 1057–1058.

———. "David Fraser Fraser-Harris, D.SC, M.D." *Proceedings of the Royal Society of Edinburgh* 57 (1938): 411–412.

———. "The Jewish Method of Slaughtering." *British Medical Journal* 1 (February 6, 1897), 345–346.

———. "The Jewish Shechita." *British Medical Journal* 2, October 21, 1893, 911.

———. Minutes of Evidence Taken Before the Committee Appointed by the Admiralty to Consider the Humane Slaughtering of Animals. London: HMSO, 1904.

———. Report of the Committee Appointed by the Admiralty to Consider the Humane Slaughtering of Animals. London: HMSO, 1904.

———. "The Slaughtering of Animals." *British Medical Journal* 1, June 9, 1894, 1256–1257.

Atkins, P.J. "The Glasgow case: meat, disease and regulation, 1889–1924." *Agricultural History Review* 52, no. 2 (2004): 161–182.

Backhouse, Constance. "Anti-Semitism and the Law in Québec City: The *Plamondon* Case, 1910–15." In *Transformations in American Legal History: Law, Ideology, and Methods, Essays in Honor of Morton J.* Horwitz, *Volume Two,* edited by Daniel W. Hamilton and Alfred Brophy, 303–325. Cambridge, MA: Harvard Law School/Harvard University Press, 2010.

Baker, K.D. "Vivisection debate in nineteenth century Great Britain: a muted echo in colonial and early post-colonial Australia." *Australian Veterinarian* 76, no. 10 (1998): 683–689.

Baldin, Damien. "De L'Horreur du Sang à l'Insoutenable Souffrance Animale: Élaboration sociale des régimes de sensibilité à la mise à mort des animaux (19e–20e siècles). *Vingtième Siècle* 123 (2014): 52–68.

Bale, Anthony and David Feldman, eds. *Blood: Reflections on What Unites and Divides Us.* Oxford: Shire, 2016.

Barber, Ben. *Jews in Glasgow, 1879–1939.* London: Vallentine Mitchell, 2007.

Bartlett, Peter. "Downer, Sir John William (1843–1915)," *Australian Dictionary of Biography.* National Centre of Biography, Australian National University, http://adb.anu.edu.au/biography/downer-sir-john-william-6007/text10261.

Beers, Diane L. *For the Prevention of Cruelty: The History and Legacy of Animal Rights Activism in the United States.* Athens: Ohio State University Press/Swallow Press, 2006.

Belkin, Simon. *Through Narrow Gates: A Review of Jewish Immigration, Colonization and Immigrant Aid Work in Canada (1840–1940).* Montreal: Canadian Jewish Congress/Jewish Colonization Association, 1966.

Benas, Bertram. "A Survey of the Jewish Institutional History of Liverpool and District." *Jewish Historical Society of England–Transactions* 17 (1951–1952): 23–37.

Bennett, John M. *Portraits of The Chief Justices of New South Wales, 1824–1977.* Sydney: John Ferguson, 1977.

———. *Sir Alfred John Stephen: Third Chief Justice of New South Wales.* Sydney: Federation Press, 2009.

Bérard, Robert Nicholas. "Moral Education in Nova Scotia." *Acadiensis* 14, no.1 (1984): 49–63.

Bergeaud-Blackler, Florence, Ari Z. Zivotofsky, and Mara Miele. "Knowledge and Attitudes of European Kosher Consumers as Revealed through Focus Groups." *Society and Animals* 21, no. 5 (2013): 425–442.

Berger, Benjamin. *Law's Religion: Religious Difference and the Claims of Constitutionalism.* Toronto: University of Toronto Press, 2015.

Bergman, G. F. J. "Davis, Alexander Barnard (1828–1913)." *Australian Dictionary of Biography.* National Centre of Biography, Australian National University, http://adb.anu.edu.au/biography/davis-alexander-barnard-3379/text5113.

Berman, Jeremiah J. *Shehitah: A Study in the Cultural and Social Life of the Jewish People*. New York: Bloch, 1941.

———. "The Trend in Religious Observance in Mid-Nineteenth Century America." *American Jewish Historical Society Publications* 37 (1947): 31–53.

Beth Israel Synagogue. *Baron De Hirsch Congregation 1890 to 1990: 100th Anniversary Commemorative Book*. Halifax, NS: Beth Israel Synagogue, 1990.

Biale, David. *Blood and Belief: The Circulation of a Symbol Between Jews and Christians*. Berkeley: University of California Press, 2008.

Black, Eugene C. *The Social Politics of Anglo-Jewry, 1880–1920*. Oxford: Basil Blackwell, 1988.

Birnbaum, Pierre and Ira Katznelson. eds. *Paths of Emancipation: Jews, States, and Citizenship*. Princeton: Princeton University Press, 1995.

Bisgould, Lesli. *Animals and the Law*. Toronto: Irwin Law, 2011.

Blaazar, David. "Sharks and Shylocks: Englishness and otherness in popular discourse on 'the City', 1870–1914." *National Identities* 18, no. 2 (2016): 139–155.

Boudreau, Michael. *City of Order: Crime and Society in Halifax, 1918–35*. Vancouver: University of British Columbia Press, 2012.

Brantz, Dorothee. "Stunning Bodies: Animal Slaughter, Judaism, and the Meaning of Humanity in Imperial Germany." *Central European History* 35, no. 2 (2002): 167–194.

Braverman, William Alan. "The ascent of Boston's Jews, 1630–1918." PhD Dissertation, Harvard University, 1990.

Breck, Allen DuPont. *The Centennial History of the Jews of Colorado, 1859–1959*. Denver: Hirschfeld Press, 1960.

Brotman, Shari L., and James L. Torczyner. *The Changing Portrait of Jewish Life in Halifax*. Halifax, NS: McGill Consortium for Ethnicity and Strategic Social Planning, 1997.

Brown, Antony. *Who Cares for Animals? 150 Years of the RSPCA*. London: Heinemann, 1974.

Brown, R. Blake and Susan S. Jones. "A Collective Biography of the Supreme Court Judiciary in Nova Scotia, 1900–2000." In *The Supreme Court of Nova Scotia 1754–2004: From Imperial Bastion to Provincial Oracle*, edited by Philip Girard, Jim Phillips, and Barry Cahill, 204–242. Toronto: University of Toronto Press/Osgoode Society for Canadian Legal History, 2004.

Brunn, S.D. ed. *The Changing World Religion Map*. Dordrecht: Springer, 2015.

Budd, Wallace B. *Hear the Other Side: A History of the Royal Society for the Prevention of Cruelty to Animals in South Australia 1875–1988*. Adelaide, SA: RSPCA, 1988.

Bushnell, Charles J. "Some Social Aspects of the Chicago Stock Yards." *American Journal of Sociology* 7, no. 2 (1901): 145–170.

Buttenwieser, Laemlein, trans. *Tub Taam or a Vindication of the Jewish Mode of Slaughtering Animals for Food Called Shechitah* (2nd ed). New York: Bloch Publishing, 1904.

Cahill, Barry. "Birth of a Lawyer: James McGregor Stewart and the Halifax Bar on the Eve of Great War." *Journal of the Royal Nova Scotia Historical Society* 3 (2000): 22–32.

Canadian Society for the Prevention of Cruelty to Animals. *Canadian Society for the Prevention of Cruelty to Animals*. Montreal: CSPCA, 1886.

Cesarani, David. *The Jewish Chronicle and Anglo-Jewry, 1841–1991*. Cambridge: Cambridge University Press, 1994.

———. "The Jews of Bristol and Liverpool, 1750–1850: Port Jewish Communities in the Shadow of Slavery." In *Jews and Port Cities 1590–1990: Commerce, Community and Cosmopolitanism*, edited by David Cesarani and Gemma Romain, 141–156. London: Vallentine Mitchell, 2006.

Cesarani, David, and Gemma Romain, eds. *Jews and Port Cities 1590–1990: Commerce, Community and Cosmopolitanism*. London: Vallentine Mitchell, 2006.

Clark, Michael. *Albion and Jerusalem: The Anglo-Jewish Community in the Post-Emancipation Era 1858–1887*. Oxford: Oxford University Press, 2009.

Claus, Peter M. "Carden, Sir Robert Walter." *Oxford Dictionary of National Biography*. http://www.oxforddnb.com.

Coleman, Sydney H. *Humane Society Leaders in America*. Albany, NY: American Humane Association, 1924.

Collins, Kenneth. "A Community on Trial: The Aberdeen Shechita Case, 1893." *Journal of Scottish Historical Studies* 30, no. 2 (2010): 75–92.

———. *Go and Learn: The International Story of Jews and Medicine in Scotland*. Aberdeen: Aberdeen University Press, 1988.

———. *Scotland's Jews: A Guide to the History and Community of the Jews of Scotland*. Glasgow: Scottish Council of Jewish Communities, 2008.

Corbin, Caroline Mala. "The Irony of *Hosanna-Tabor Evangelical Lutheran Church and School v. EEOC*." *Northwestern University Law Review Colloquy* 106, no.2 (2011): 96–110.

Creco, Clara. "The Superior Court Judiciary of Nova Scotia, 1754–1900: A Collective Biography." In *Essays in the History of Canadian Law*, edited by Philip Girard and Jim Phillips, 42–79. Toronto: University of Toronto Press/Osgoode Society for Canadian Legal History, 1990.

Daley, Timothy T. "A History of the Nova Scotia Family Court." in Kimball, Judge R. E., *The Bench: The History of Nova Scotia's Provincial Courts*. Halifax: Province of Nova Scotia, 1989: 143–150.

Damade, Jacques. *Abattoirs de Chicago*. Paris: Éditions La Bibliothèque, 2016.

Davis, Janet M. *The Gospel of Kindness: Animal Welfare and the Making of Modern America*. Oxford: Oxford University Press, 2016.

Day, Jared N. "Butchers, Tanners, and Tallow Chandlers: The Geography of Slaughtering in Early-Nineteenth-Century New York City." In *Meat, Modernity, and the Rise of the Slaughterhouse*, edited by Paula Young Lee, 178–197. Hanover, NH: University Press of New England, 2008.

Dembo, J. A. *The Jewish Method of Slaughter Compared with Other Methods from the Humanitarian, Hygienic, and Economic Points of View*. London, 1894.

Denoyelles, Adrienne. "'Peculiar Resistance': Tuberculosis, Identity and Conflict among Jewish Physicians in Early-Twentieth-Century America." *American Jewish History* 100, no. 3 (2016): 349–377.

Duschinsky, Charles. "Jacob Kimchi and Shalom Buzaglo." *Jewish Historical Society of England–Transactions* 7 (1911): 272–290.

Dwyer-Ryan, Meaghan. "Ethnic Patriotism: Boston's Irish and Jewish Communities, 1880–1929." PhD Dissertation, Boston College, 2010.

Edelman, Todd. *The Jews in Britain, 1656 to 2000*. Berkeley: University of California Press, 2000.

Efron, John M. "The Most Cruel Cut of All? The Campaign Against Jewish Ritual Slaughter in Fin-de-Siècle Switzerland and Germany." *Leo Baeck Institute Year Book* 52 (2007): 167–184.

Eisenberg, Ellen. *Jewish Agricultural Colonies in New Jersey, 1882–1920*. Syracuse: Syracuse University Press, 1995.

Eisgruber, Christopher L. and Lawrence G. Sager. "The Vulnerability of Conscience: The Constitutional Basis for Protecting Religious Conduct." *University of Chicago Law Review* 61, no. 4 (1994): 1245–1315.

Eker, Glen, and Deborah Pekilis. *Index of Jews Resident in New Brunswick, Nova Scotia and Prince Edward Island According to the 1861 to 1901 Censuses of Canada*. Toronto: Ontario Genealogical Society, 2004.

Elliott, Jean Leonard, ed. *Immigrant Groups*. Scarborough, ON: Prentice Hall, 1971.

Elliott, Shirley B., ed. *The Legislative Assembly of Nova Scotia, 1758–1983: A Biographical Directory*. Halifax: Province of Nova Scotia, 1984.

Elton, Benjamin J. *Britain's Chief Rabbis and the Religious Character of Anglo-Jewry, 1880–1970*. Manchester: Manchester University Press, 2000.

Emanuel, Charles H.L. *A Century and a Half of Jewish History: Extracted from the Minute Books of the London Committee of Deputies of the British Jews*. London: Routledge, 1901.

Encel, Sol. "Antisemitism and Prejudice in Australia." *Patterns of Prejudice* 23, no. 1 (1989): 16–27.

Fairfax, J. O. "Fairfax, John (1804–1877)," *Australian Dictionary of Biography*. National Centre of Biography, Australian National University, http://adb.anu.edu.au/biography/fairfax-john-3493/text5357.

Favre, David, and Vivien Tsang. "The Development of Anti-Cruelty Laws During the 1800s." *Detroit College of Law Law Review* 1 (1993): 1–35.

Figler, Bernard. *Biography of Sam Jacobs*. Gardenvale, PQ: Harpell's Press Cooperative, 1959 and 1970.

———. *Rabbi Dr. Herman Abramowitz, Lazarus Cohen, Lyon Cohen*. Gardenvale, PQ: Harpell's Press Cooperative, 1968.

Finestein, Israel. *Anglo-Jewry in Changing Times*. London: Vallentine Mitchell, 1999.

Fingard, Judith. *The Dark Side of Life in Victorian Halifax*. Porters Lake, NS: Pottersfield Press, 1989.

———. "The Prevention of Cruelty, Marriage Breakdown and the Rights of Wives in Nova Scotia, 1880-1900." *Acadiensis* 22, no. 2 (1993): 84–101.

———. "Race and Respectability in Victorian Halifax." *Journal of Imperial and Commonwealth History* 20, no. 2 (1992): 169–195.

Forbes, E.R. "Prohibition and the Social Gospel in Nova Scotia," *Acadiensis* 1, no. 1 (1971): 11–36.

Fraser, David. "The Blood Libel in North America: Jews, Law, and Citizenship in the Early 20th Century." *Law and Literature* 28, no. 1 (2016): 33–85.

———. *"Honorary Protestants:" The Jewish School Question in Montreal, 1867–1997.* Toronto: University of Toronto Press/Osgoode Society for Canadian Legal History, 2015.

Friedenberg, Albert M. "The Year 5669." *American Jewish Year Book* 11 (1909–1910): 55–102.

Friedenreich, David M. *Foreigners and Their Food: Constructing Otherness in Jewish, Christian, and Islamic Law.* Berkeley: University of California Press, 2011.

Gaudreault-DesBiens, Jean-François, ed. *Le Droit, La Religion et le "Raisonnable."* Montreal: Thémis, 2009.

Getzler, Israel. *Neither Toleration nor Favour: The Australian Chapter of Jewish Emancipation.* Melbourne, VIC: Melbourne University Press, 1970.

Gilfillan, Mark. "Jews in Edinburgh: 1880–1950." PhD dissertation, University of Ulster, 2012.

Gillis, Ronald, and Paul C. Whitehead. "The Halifax Jews: A Community within a Community." In *Immigrant Groups*, edited by Jean Leonard Elliott, 89–94, Scarborough, ON: Prentice Hall, 1971.

Gilham, Jamie. *Loyal Enemies: British Converts to Islam, 1850–1950.* London: Hurst, 2014.

———. "'Upholding the Banner of Islam:' British Converts to Islam and the Liverpool Muslim Institute, c.1887–1908." *Immigrants and Minorities* 33, no. 1 (2015): 23–44.

Gilman, Sander L "Kafka Wept." *Modernism/Modernity* 1, no.1 (1994): 17–37.

Girard, Philip. "The Roots of a Professional Renaissance: Lawyers in Nova Scotia 1850–1910." *Manitoba Law Journal* 20, no. 1 (1991): 148–180.

———. "The Supreme Court of Nova Scotia, Responsible Government, and the Quest of Legitimacy, 1850–1920." *Dalhousie Law Journal* 17, no. 2 (1994): 430–457.

Girard, Philip, Jim Phillips, and Barry Cahill, eds. *The Supreme Court of Nova Scotia, 1754–2004: From Imperial Bastion to Provincial Oracle.* Toronto: University of Toronto Press/Osgoode Society for Canadian Legal History, 2004.

Girard, Philip and Jim Phillips, eds. *Essays in the History of Canadian Law.* Toronto: University of Toronto Press/Osgoode Society for Canadian Legal History, 1990.

Glover, David. *Literature, Immigration, and Diaspora in Fin-de-Siècle England: A Cultural History of the 1905 Aliens Act.* Cambridge: Cambridge University Press, 2012.

Godfrey, Sheldon J., and Judith C. Godfrey. *Search Out the Land: The Jews and the Growth of Equality in British Colonial America, 1740–1867.* Montreal: McGill-Queen's University Press, 1995.

Goeb, Jan. *History of the Jews in the Maritimes.* Undated typescript, Jewish Historical Society Fonds, MG 20-Vol. 561-7. Nova Scotia Archives, Halifax.

Goldberg, P. Selvin. *The Manchester Congregation of British Jews, 1857–1957.* Manchester: Congregation of British Jews, 1957.

Goodman, Mervyn. "The Jewish Community of Liverpool." *Jewish Journal of Sociology* 38, no. 2 (1996): 89–104.

Goodwin, George M., and Ellen Smith, eds. *The Jews of Rhode Island*. Hanover, NH: University Press of New England, 2004.

Gordon, Jerry. *From Its Beginnings: The Jewish Community of the Atlantic City Area*. Atlantic City: published by author, 2016.

Goren, Arthur A. *New York Jews and the Quest for Community: The Kehillah Experiment, 1908–1922*. New York: Columbia University Press, 1970.

Grandin, Temple. "Improving religious slaughter practices in the U.S." *Anthropology of food* [Online], 5, May 2006, http://aof.revues.org/93 2006.

———. "Problems with kosher slaughter." *International Journal for the Study of Animal Problems* 1 (1980): 375–390.

Grandin, Temple and Joe M. Regenstein. "Religious Slaughter and Animal Welfare: a Discussion for Meat Scientists." *Meat Focus International* 1994: 115–123.

Greg, I.M., and S.H. Towers. *Cattle ships, and our meat supply*. London, 1894.

Gross, Aaron S. *The Question of the Animal and Religion: Theoretical Stakes, Practical Implications*. New York: Columbia University Press, 2015.

Grunfeld, I. *The Religious and Moral Basis of the Jewish Dietary Laws*. London: National Council of Shechita Boards, 1966.

Gurock, Jeffrey S. *Orthodox Jews in America*. Bloomington: Indiana University Press, 2009.

Hamilton, Daniel W. and Alfred Brophy, eds. *Transformations in American Legal History: Law, Ideology, and Methods, Essays in Honor of Morton J.* Horwitz, *Volume Two*. Cambridge, MA: Harvard Law School/Harvard University Press, 2010.

Harrison, Brian. "Religion and Recreation in Nineteenth-Century England." *Past and Present* 38, no. 1 (1967): 98–125.

Hart, Arthur Daniel, ed. *The Jew in Canada*. Toronto: Jewish Publications,1926.

Hart, Mitchell B. *The Healthy Jew*. Cambridge: Cambridge University Press, 2007.

Henriques, H.S.Q. *The Jews and the English Law*. Oxford: Oxford University Press, 1908.

———. *The Return of the Jews to England: Being a Chapter in the History of English Law*. London: Macmillan, 1905.

Herman, Didi. *An Unfortunate Coincidence: Jews, Jewishness, and English Law*. Oxford: Oxford University Press, 2011.

Hirschl, Ran and Ayelet Shachar. "The Constitutional Boundaries of Religious Accommodation." In *Constitutional Secularism in an Age of Religious Revival*, edited by Susanna Mancini and Michael Rosenfeld, 175–191. Oxford: Oxford University Press, 2014.

Hoberman, Michael. *How Strange It Seems: The Cultural Life of Jews in Small-Town New England*. Amherst: University of Massachusetts Press, 2008.

Homa, Bernard. *Shehita*. London: Soncino Press/Board of Deputies of British Jews, 1967.

Horowitz, Roger. *Kosher USA: How Coke Became Kosher and Other Tales of Modern Food*. New York: Columbia University Press, 2016.

Horvitz, Eleanor F., and Geraldine S. Foster. "Jewish Farmers in Rhode Island and Nearby Massachusetts." *Rhode Island Jewish Historical Notes* 10, no. 4 (1990): 443–478.

Hughes, Gordon and Claire Lawson. "RSPCA and the criminology of social control." *Crime, Law and Social Change*, 55, no. 5 (2011): 375–389.

Hyamson, Albert M. *David Salomons*. London: Methuen, 1939.

———. *A History of the Jews of England*. London: Jewish Historical Society / Chatto & Windus, 1908.

———. *The London Board for Shechita, 1804–1954*. London: London Board for Shechita, 1954.

———. *The Sephardim of England: A History of the Spanish and Portuguese Jewish Community, 1492–1951*. London: Methuen, 1951.

Hyamson, M. "Another Word on Jewish Dietary Laws." *Jewish Quarterly Review* 9, no. 2 (1897): 294–310.

Hyman, Moses. "The Jewish Method of Slaying Animals: From the Point of View of Humanity." *American Jewish Year Book* 25 (1924): 160–179.

Iannacone, Stephen. "Felony Animal Cruelty Laws in New York." *Pace Law Review* 31, no. 2 (2011): 748–770.

Jewish Social Services Greater Manchester. *They Came from the Haim: A History of Manchester Jewry from 1867*. Manchester: Jewish Social Services Greater Manchester, 1995.

Johnston, Beatrice. *For Those Who Cannot Speak: A History of the Canadian Society for the Prevention of Cruelty to Animals, 1869–1969*. Chomedy, PQ: Dev-Sco, 1970.

Joseph, Anna. "Going Dutch: A Model for Reconciling Animal Slaughter Reform with the Religious Freedom Restoration Act." *Journal of Animal Ethics* 6, no. 2 (2016): 133–152.

Joseph, Samuel. *History of the Baron de Hirsch Fund: The Americanization of the Jewish Immigrant*. Philadelphia: Jewish Publication Society, 1935.

Judd, Robin. *Contested Rituals: Circumcision, Kosher Butchering, and Jewish Political Life in Germany, 1843–1933*. Ithaca, NY: Cornell University Press, 2007.

Julius, Anthony. *Trials of the Diaspora: A History of Anti-Semitism in England*. Oxford: Oxford University Press, 2010.

Katz, Jacob. *Out of the Ghetto: The Social Background of Jewish Emancipation, 1770–1870*. Syracuse: Syracuse University Press, 1998.

Kaye, Ronald. "The Politics of Religious Slaughter: Strategies for Ethno-religious Political Action." *New Community* 19, no. 2 (1993): 235–250.

Kean, Hilda. *Animal Rights: Political and Social Change in Britain since 1800*. London: Reaktion, 1998.

Kenefick, William. "The Jews and Irish in Modern Scotland: Anti-Semitism, Sectarianism and Social Mobility." *Immigrants and Minorities* 31, no. 2 (2013): 189–213.

Kim, Claire Jean. *Dangerous Crossings: Race, Species, and Nature in a Multicultural Age*. Cambridge: Cambridge University Press, 2015.

Kimball, Judge R. E. *The Bench: The History of Nova Scotia's Provincial Courts*. Halifax: Province of Nova Scotia, 1989.

Klug, Brian. "Ritual Murmur: The Undercurrent of Protest Against Religious Slaughter of Animals in Britain in the 1980s." *Patterns of Prejudice* 23, no. 2 (1989): 16–28.

Kokosalakis, N. *Ethnic Identity and Religion: Transition and Change in the Liverpool Jewry*. Washington, DC: University Press of America, 1982.

Kosmin, Barry A. "Localism and pluralism in British Jewry, 1900–80." *Jewish Historical Society of England–Transactions* 28 (1981–1982): 111–125.

Kraemer, David C. *Jewish Eating and Identity Through the Ages*. New York: Routledge, 2007.

Kushner, Tony. *Anglo-Jewry Since 1066: Place, Locality, and Memory*. Manchester: Manchester University Press, 2009.

Kymlicka, Will and Sue Donaldson. "Animal Rights, Multiculturalism, and the Left." *Journal of Social Philosophy* 45, no. 1 (2014): 116–145.

Lafferty, Renee. *The Guardianship of Best Interests: Institutional Care for the Children of the Poor in Halifax, 1850–1960*. Montreal: McGill-Queen's University Press, 2013.

Langham, Raphael. *250 Years of Convention and Contention: A History of the Board of Deputies of British Jews, 1760–2010*. London: Vallentine Mitchell, 2010.

Lawrence, David M. "Private Exercise of Governmental Power." *Indiana Law Journal* 61, no. 4 (1986): 647–695

Lederhendler, Eli. *American Jewry: A New History*. Cambridge: Cambridge University Press, 2017.

Lee, Paula Young, ed. *Meat, Modernity, and the Rise of the Slaughterhouse*. Hanover NH: Press of New England, 2008.

Levey, Geoffrey Brahm, and Philip Mendes, eds. *Jews and Australian Politics*. Brighton: Sussex Academic Press, 2004.

Levin, S.I., and Edward A. Boyden. *The Kosher Code of the Orthodox Jew*. Minneapolis: University of Minnesota Press, 1940.

Levy, A. *The Origins of Glasgow Jewry, 1812-1895*. Glasgow: printed by author, 1949.

———. "The Origins of Scottish Jewry." *Jewish Historical Society of England– Transactions*, 19 (1955): 129–163.

Li, Chien-hui. "Mobilizing Christianity in the Antivivisection Movement in Victorian Britain." *Journal of Animal Ethics* 2, no. 2 (2012): 141–161.

———. "A Union of Christianity, Humanity, and Philanthropy: The Christian Tradition and the Prevention of Cruelty to Animals in Nineteenth-Century England." *Society and Animals* 8, No. 1 (2000): 265–285.

Liedtke, Rainer. *Jewish Welfare in Hamburg and Manchester c. 1850–1914*. Oxford: Clarendon Press, 1998.

Lipman, V.D. "A Survey of Anglo-Jewry in 1851." *Jewish Historical Society of England– Transactions* 17 (1951–1952): 171–188.

London Board of Shechita: *Laws of the Board for Administering the Affairs of* שיחטה. London, 1880.

London Committee of Deputies of British Jews. 54th *Annual Report*. London: London Committee of Deputies of British Jews, 1904.

Lytton, Timothy D. *Kosher: Private Regulation in the Age of Industrial Food*. Cambridge, MA: Harvard University Press, 2013.

McConnell, Anita. "Laurie, Sir Peter." *Oxford Dictionary of National Biography*. http://www.oxforddnb.com.

MacCormac, Henry. *The Painless Extinction of Life in Animals Designed for Human Food*. London, 1864.

MacLachlan, Ian. "*Coup de Grâce:* Humane Cattle Slaughter in Nineteenth Century Britain." *Food and History* 3, no. 2 (2005): 145–171.

———. "Humanitarian Reform, Slaughter Technology, and Butcher Resistance in Nineteenth-Century Britain." In *Modernity, and the Rise of the Slaughterhouse*, edited by Paula Young Lee, 107–112. Hanover, NH: University Press of New England, 2008.

Mancini, Susanna and Michel Rosenfeld, eds. *Constitutional Secularism in an Age of Religious Revival*. Oxford: Oxford University Press, 2014.

Manning, Aubrey and James Serpell, eds. *Animals and Human Society: Changing Perspectives*. New York: Routledge, 1994.

Marcus, Jacob Rader. *Early American Jewry: The Jews of New York, New England and Canada, 1649–1794*. Philadelphia: Jewish Publication Society, 1961.

Marcus, Jacob Rader and Judith M. Daniels, eds. *The Concise Dictionary of American Jewish Biography*. Brooklyn: Carlson Publishers, 1994.

Marquis, Greg. "The History of Policing in the Maritime Provinces." *Urban History Review* 19, no. 2 (1990): 84–99.

McConnell, Michael W. "Free Exercise Revisionism and the *Smith* Decision." *University of Chicago Law Review* 57, no. 4 (1990): 1109–1153.

McManus, Phil. "Feeding a growing city: The Glebe Island Abattoir and the provision of meat for Sydney." *Rural Society* 1, no. 3 (2001): 243–253.

McCrea, Roswell C. *The Humane Movement: A Descriptive Survey*. New York: Columbia University Press, 1910.

Medjuck, Sheva. *Jews of Atlantic Canada*. St. John's, NL: Breakwater, 1986.

Memmi, Albert, W. Ackermann, and S. Zoberman. "Pratique Religieuse et Identité Juive." *Revue française de sociologie* 14 (1973): 242–270.

Mermelstein, Ari, Victoria Saker Woeste, Ethan Zadoff, and Marc Galanter. *Jews and the Law*. New Orleans: Quid Pro Books, 2014.

Mesmer, Beatrix. "The Banning of Ritual Slaughter in Switzerland." *Leo Baeck Institute Year Book* 52 (2007): 185–194.

Metcalfe, Robyn S. *Meat, Commerce and the City: The London Food Market, 1800-1855*. London: Pickering and Chatto, 2012.

Mew, James, and Nilanjana Banerji. "Bodkin, Sir William Henry." *Oxford Dictionary of National Biography*. http://www.oxforddnb.com.

Miller, Rev. M.R. *The Luminous Unity*. Philadelphia, 1874.

Montefiore, C.G. "Dr. Wiener on the Dietary Laws." *Jewish Quarterly Review* 8, no. 3 (1896): 392–413.

Moon, Richard, ed. *Law and Religious Pluralism in Canada*. Vancouver: University of British Columbia Press, 2008.

Moss, Arthur W. *Valiant Crusade: The History of the Royal Society for the Prevention of Cruelty to Animals*. London: Cassell, 1961.

Munk, Michael L., and Eli Munk eds. *Shechita: Part II of "Edut Ne'emana: Religious and Historical Research on the Jewish Method of Slaughter*. Brooklyn: Gur Aryeh, 1976.

Murphy, Terrence. "The Religious History of Atlantic Canada: The State of the Art." *Acadiensis* 15, No. 1 (1985): 152–174.

Newman, Aubrey, ed. *Provincial Jewry in Victorian England*. London: Jewish Historical Society of England, 1975.

———. "Provincial Jewry in Victorian Britain: A Report." *Jewish Historical Society of England–Transactions* 25 (1973–1975): 222–229.

Nizard-Benchimol, Sophie. "L'abattage dans la tradition juive." *Études rurales* 147–148 (1998): 49–64.

Nizard, Sophie, Séverine Mathieu, Aïda Kanafani-Zahar, eds. *A Croire et à Manger: Religions et Alimentation*. Paris: L'Harmattan, 2008.

Normand, Sylvio. "L'*affaire* Plamondon: un cas d'antisémitisme à Québec au début du XXe siècle." Cahiers de Droit 48, no 3 (2007): 477–504.

Nurse, Angus. "Beyond the Property Debate: Animal Welfare as a Public Good." *Contemporary Justice Review* 19, no. 2 (2016): 174–187.

Oppenheim, Samson D. "The Jewish Population of the United States." *American Jewish Yearbook* 20 (1918–1919): 31–74.

Oppenheim, Samuel. "The Question of Kosher Meat Supply in New York in 1813: With a Sketch of Earlier Conditions." *Publications of the American Jewish Historical Society* 25 (1917): 31–62.

Otter, Chris. "Civilizing Slaughter: The Development of the British Public Abattoir, 1850–1910." In *Meat, Modernity, and the Rise of the Slaughterhouse*, edited by Paula Young Lee, 89–106. Hanover NH: University Press of New England, 2008.

Pearson, Susan J. *The Rights of the Defenseless: Protecting Animals and Children in Gilded Age America*. Chicago: University of Chicago Press, 2011.

Pfeffer, Jeremy I. *'From One End of the Earth to the Other': The London Bet Din, 1805–1855, and the Jewish Convicts Transported to Australia*. Brighton: Sussex Academic Press, 2008.

Phillips, N. Taylor. "Family History of the Reverend David Mendez Machado." *Publications of the American Jewish Historical Society* 2 (1894): 45–61.

Polland, Anne and Daniel Soyer. *Emerging Metropolis: New York Jews in the Age of Immigration, 1840–1920*. New York: New York University Press, 2012.

Popper, Deborah E. "'Great Opportunities for Many of Small Means:' New Jersey's Agricultural Colonies." *Geographical Review* 96, no. 1 (2006): 24–49.

Porter, Muriel. *The Argus: The Life and Death of a Great Melbourne Newspaper (1846–1957)*. Melbourne, VIC: RMIT Press, 2003.

Potter, Simon. "Communication and Integration: The British and Dominions Press and the British World, c. 1876–1914." *Journal of Imperial and Commonwealth History* 31, no. 2 (2003): 190–206.

Poulter, Sebastian. *English Law and Ethnic Minority Customs*. London: Butterworths, 1986.

———. "Ethnic Minority Customs, English Law and Human Rights." *International and Comparative Law Quarterly* 36, no. 3 (1987): 589–615.

Pozzi, P.S., W. Geraisy, S. Barakeh, and M. Azaran. "Principles of Jewish and Islamic Slaughter with Respect to OIE (World Organization of Animal Health) Recommendations." *Israel Journal of Veterinary Medicine* 70, no. 3 (2015): 3–16.

Pringle, A.S. "The Verdict of "Not Proven" in Scotland," *Canadian Law Review* 4 (1905): 248–256.

Rechcigl, Miloslav Jr. *Encyclopedia of Bohemian and Czech-American Biography*, Vol. 1. Bloomington, IN: AuthorHouse, 2016.

Renteln, Alison Dundes. *The Cultural Defense*. Oxford: Oxford University Press, 2004.

Rischin, Moses and John Livingston. *Jews of the American West*. Detroit: Wayne State University Press, 1991.

Ritvo, Harriet. *The Animal Estate: The English and Other Creatures in the Victorian Age*. London: Penguin, 1990.

———. "Animals in Nineteenth-Century Britain: Complicated Attitudes and Competing Categories." In *Animals and Human Society: Changing Perspectives*, edited by Aubrey Manning and James Serpell, 106–126. New York: Routledge, 1994.

Robinson, Ira and Mervin Butovsky. *Renewing Our Days: Montreal Jews in the Twentieth Century*. Montreal: Véhicule Press, 1995.

Robinson, Ira. "The Bouchard-Taylor Commission and the Jewish Community of Québec in Historical Perspective." In *Religion, Culture, and the State*, edited by Howard Adelman and Pierre Anctil, 58–68. Toronto: University of Toronto Press, 2011.

———. *A History of Antisemitism in Canada*. Waterloo, ON: Wilfrid Laurier University Press, 2015.

———. *Rabbis and Their Community: Studies in the Eastern European Orthodox Rabbinate in Montreal, 1896–1930*. Calgary, AB: University of Calgary Press, 2007.

Rock, Howard B. *Haven of Liberty: New York Jews in the New World, 1654–1865*. New York: New York University Press, 2012.

Rome, David. *The Plamondon Case and S.W. Jacobs*, 2 vols. Montreal: Canadian Jewish Congress National Archives, 1982.

———. *The Jewish Biography of Henri Bourassa, Part 1*. Montreal: National Archives of the Canadian Jewish Congress, 1988.

Rosenberg, Louis. *Canada's Jews: A Social and Economic Study of Jews in Canada in the 1930s*. Edited by Morton Weinfeld. Montreal: McGill-Queen's University Press, 1993.

Rosenberg, Louise. "Boas, Abraham Tobias (1842–1923)." *Australian Dictionary of Biography*. National Centre of Biography, Australian National University, http://adb.anu.edu.au/biography/boas-abraham-tobias-5277/text8897.

———. "Rev. A.T. Boas: A Pioneer Jewish Minister." *Australian Jewish History Society Journal* 7, no. 2 (1972): 77–169.

Rosenberg, Stuart. *The Canadian Jewish Community*, Vol.1. Toronto: McClelland and Stewart, 1970.

Roth, Cecil. *The Rise of Provincial Jewry: The Early History of the Jewish Communities in the English Countryside, 1740–1840*. London: Jewish Monthly, 1950.

Rowley, Francis H. *The Humane Idea*. Boston: American Humane Education Society, 1912.

———. *Slaughter-House Reform in the United States and the Opposing Forces*. Boston: Massachusetts Society for the Prevention of Cruelty to Animals, c. 1913.

Royal Society for the Prevention of Cruelty to Animals. *Ninetieth Annual Report, 1913*. London: RSPCA, 1913.

Rubenstein, Hilary. *Chosen: The Jews in Australia*. Sydney: Allen & Unwin, 1987.

———. *The Jews in Australia: A Thematic History, Volume One, 1788–1945*. Port Melbourne, VIC: William Heinemann Australia, 1991.

Rubinstein, W.D. *The Jews in Australia*. Melbourne, VIC: Australasian Educa, 1986.

Rubinoff, Michael W. "Rabbi in a Progressive Era: C. E H. Kauvar of Denver." *Colorado Magazine* 54, no. 3 (1977): 220–239.

Rutland, Suzanne D. *Edge of the Diaspora: Two Centuries of Jewish Settlement in Australia*. Sydney: Collins, 1988.

———. *The Jews in Australia*. Cambridge: Cambridge University Press. 2005.

Rutledge, Martha. "Stephen, Sir Alfred (1802–1894)." *Australian Dictionary of Biography*. National Centre of Biography, Australian National University, http://adb.anu.edu.au/biography/stephen-sir-alfred-1291/text7645.

Sack, B.G. *Canadian Jews-Early in This Century*. Montreal: Canadian Jewish Congress National Archives, 1975.

Salbstein, M.C.N. *The Emancipation of the Jews in Britain: The Question of the Admission of the Jews to Parliament, 1828–1860*. Rutherford, NJ: Fairleigh Dickinson University Press, 1982.

Sarna, Jonathan D. *American Judaism: A History*. New Haven, CT: Yale University Press, 2004.

Sarna, Jonathan, Ellen Smith and Scott-Martin Kosofsky, eds. *The Jews of Boston*. New Haven: Yale University Press, 2005.

Sassoon, Rabbi Solomon David. *A Critical Study of Electrical Stunning and the Jewish Method of Slaughter (Shechita)*. Letchford, Herts: published by author, 1955.

Scott, Colin. "Private Regulation of the Public Sector: A Neglected Facet of Contemporary Governance." *Journal of Law and Society* 29, no. 1 (2002): 56–76.

Seccombe, Thomas, and H.C.G. Mathew. "Ballantine, William." *Oxford Dictionary of National Biography*. http://www.oxforddnb.com.

Segal, Beryl. "Congregation Sons of Zion Providence Rhode Island: The Orms Street Synagogue." *Rhode Island Jewish Historical Notes* 4, no. 3 (1965): 239–277.

Sewell, Anna. *Black Beauty: His Grooms and Companions*. Edited by Kristen Guest. Newcastle: Cambridge Scholars Publishing, 2011.

Shargel, Baila R. "Louis Ginzberg As Apologist." *Jewish History* 79, no. 2 (1989): 210–220.

Sherman, Moshe D. "Struggle for Legitimacy: The Orthodox Rabbinate in Mid-Nineteenth Century America." *Jewish History* 10, no. 1 (1996): 63–74.

Simons, Hyman A. "Chief Rabbi Dr. Hermann Adler– A Period Study in Anglo-Jewry." *Judaism* 18, no. 2 (1969): 223–231.

Simpson, W.J. Report to the Public Health Committee on the Inspection of Meat and the Need of Public Abattoirs in Aberdeen. Aberdeen, 1885.

Singer, Steven. "Jewish Religious Thought in Early Victorian London." *Association of Jewish Studies Review* 10, no. 2 (1985): 181–210.

Smith, Judith E. *Family Connections: A History of Italian and Jewish Immigrant Lives in Providence Rhode Island, 1900-1940*. Albany, NY: State University of New York Press, 1985.

Snyder, Holly. "Reconstructing the Lives of Newport's Hidden Jews, 1740–1790." In *The Jews of Rhode Island,* edited by George M. Goodwin and Ellen Smith, 27–39. Hanover, NH: University Press of New England, 2004.

Stephen, Sir Alfred. *Australian Divorce Bills: The Objections to Them, Religious and Social, Considered.* Sydney, 1888.

Sutherland, David A. "Race Relations in Halifax, Nova Scotia, During the Mid-Victorian Quest for Reform." *Journal of the Canadian Historical Association* 7, no. 1 (1996): 35–54.

———. "Voluntary Societies and the Process of Middle-class Formation in Early Victorian Nova Scotia." *Journal of the Canadian Historical Association* 5, no. 1 (1994): 237–263.

Thomas, Courtney I.P. *In Food We Trust: The Politics of Purity in American Food Regulation.* Lincoln: University of Nebraska Press, 2014.

Tischler, Joyce. "The History of Animal Law, Part I (1972–1987)." *Stanford Journal of Animal Law and Policy* 1, no. 1 (2008): 1–49.

Trabsky, Marc. "Institutionalising The Public Abattoir in Nineteenth Century Colonial Society." *Australian Feminist Law Journal* 40, no. 2 (2014): 169–184.

Tulchinsky, Gerald. *Taking Root: The Origins of the Canadian Jewish Community*. Hanover, NH: University Press of New England, 1993.

Turner, James. *Reckoning with the Beast: Animals, Pain, and Humanity in the Victorian Mind.* Baltimore: Johns Hopkins University Press, 1981.

Uchill, Ida Libert. *Pioneers, Peddlers and Tsadikim: The Story of Jews in Colorado*. Boulder: University Press of Colorado, 2000.

Usher, Jim. *The Argus: The Life and Death of a Newspaper.* Melbourne, VIC: Australian Scholarly Publishing, 2007.

Van Praagh, Shauna. "View from the *Succah:* Religion and Neighbourly Relations." In *Law and Religious Pluralism in Canada,* edited by Richard Moon, 21–40. Vancouver: University of British Columbia Press, 2008.

Vaughan, Laura and Alan Penn. "Jewish Immigrant Settlement Patterns in Manchester and Leeds 1881." *Urban Studies* 43, no. 3 (2006): 653–671.

Vernon, Leonard F., and Allen Meyers. *Jewish South Jersey*. Charleston, SC: Arcadia, 2007.

Vialles, Noëlie. "Légitimités et Légalités: À Propos des Viandes Rituelles." In *A Croire et à manger: Religions et alimentation,* edited by Sophie Nizard, Séverine Mathieu, Aïda Kanafani-Zahar, 209–231. Paris: L'Harmattan, 2008.

Wachtel, David. "Jewish Burial Societies: The Origins and Development of the *Hevra Kaddisha*." *Jurist* 59 (1999): 214–228.

Waterman, Stanley. "Eating, Drinking and Maintenance of Community: Jewish Dietary Laws and Their Effects on Separateness." In *The Changing World Religion Map*, edited by Stanley D. Brunn, 2867–2880. Dordrecht: Springer, 2015.

Weinbren, Dan. "Against All Cruelty: The Humanitarian League." *History Workshop* 38, no. 1 (1994): 86–105.

Weintraub, Irwin. *Jews in American Agriculture: An Annotated Bibliography.* Jefferson NC: McFarland, 2007.

Weisberg, Richard H., ed. "A Thousand Years of Infamy: The History of the Blood Libel." *Law and Literature* 28, no.1 (2016).

Weissbach, Lee Shai. *Jewish Life in Small-Town America: A History*. New Haven: Yale University Press, 2005.

Wiener, A. *Die Jüdischen Speisegesetze*. Breslau, 1895.

Williams, Bill. *Jewish Manchester: An Illustrated History*. Derby: Breedon, 2008.

———. *The Making of Manchester Jewry, 1740–1875*. Manchester: Manchester University Press, 1985.

Wyszkowski, Yehezkel. "*The American Hebrew:* An Exercise in Ambivalence." *American Jewish History* 76, no. 3 (1987): 340–353.

Zellick, Graham. "Freedom of Religion and the Jewish Community in the United Kingdom." *Patterns of Prejudice* 21, no. 2 (1987): 3–16.

Index

Note: Page numbers followed by 'n' refer to notes

www.ingramcontent.com/pod-product-compliance
Ingram Content Group UK Ltd.
Pitfield, Milton Keynes, MK11 3LW, UK
UKHW021920010526
470548UK00007B/61

9 781618 117427